Construction Scheduling
with Primavera Enterprise

Second Edition

Construction Scheduling with Primavera Enterprise

Second Edition

**David A. Marchman
and
Terry L. Anderson**

THOMSON

DELMAR LEARNING   Australia Canada Mexico Singapore Spain United Kingdom United States

THOMSON

DELMAR LEARNING

Construction Scheduling with Primavera Enterprise, Second Edition

David A. Marchman and Terry L. Anderson

Business Unit Director:
Alar Elken

Executive Editor:
Sandy Clark

Acquisitions Editor:
Mark Huth

Development:
Dawn Daughterty

Executive Marketing Manager:
Maura Theriault

Channel Manager:
Fair Huntoon

Marketing Coordinator:
Brian McGrath

Executive Production Manager:
Mary Ellen Black

Production Manager:
Larry Main

Production Editor:
Stacy Masucci

Library of Congress Cataloging-in-Publication Data:

0-7668-2861-1

NOTICE TO THE READER

Contents

Preface

ABOUT THIS BOOK

This book is a graphic, step-by-step introduction to good construction project control techniques using *Primavera Project Planner for the Enterprise (P3e)*. It covers the traditional theory on planning, scheduling, and controlling construction projects. Topics included are schedule development, activity definition, relationships, calculations, resources, costs, and monitoring, documenting, and controlling change. An important difference between this and other scheduling texts is that this book takes the student through all the above topics with an example construction project schedule using *P3e*. The student is not only exposed to on-screen images but is also shown how to manipulate hard-copy prints for exceptional communication and demonstration results.

This book has four strengths. First, you do not have to start at the beginning of the book and go to the end to derive use from it. The detail provided in the Contents makes it easy to find topic information. Second, detailed examples (sample schedules) are used to show the various scheduling information. Third, this book can be used in conjunction with *P3e's* Help Topics and tutorial for an efficient introduction to *P3e*. Last, a sample problem and exercises at the end of each chapter help to make this text very useful to learners in both classroom and industry settings. Anyone who completes the exercises by using *P3e* will have a sound basis for producing construction schedules.

COMPUTERS IN CONSTRUCTION SCHEDULING

The Advantages of Using Computers in Construction Scheduling

One of the primary advantages of using computers for construction scheduling is that the mathematical computations are instantaneous and error free. The speed and accuracy of the mathematical scheduling computations and the analysis of the information produced make computerized scheduling an invaluable tool for construction project controls.

The Advantages of *P3e*

P3e offers all the advantages of the use of computers for construction scheduling. The superb on-screen and hard-copy graphics of *P3e* make it an effective communications and project controls tool for owners, contractors, subcontractors, suppliers, and vendors. The advantages of *P3e,* compared to other scheduling software programs, are:

- *P3e* is a comprehensive, multiproject planning and control software.
- *P3e* is built on *Oracle* and *Microsoft SQL Server* relational databases for easy transfer and scalability of project management information.
- *P3e* can be used independently for project and resource management, or it can be used in conjunction with other software products.

USING THIS BOOK

This book can be used either as a stand-alone text or in conjunction with another more traditional planning/scheduling book. This book is designed primarily as a problems-oriented lab manual to improve student usage of *P3e*. Students will be able to use this book as a step-by-step guide while they are sitting at their computer screen.

Although this book is intended for the classroom, the step-by-step tutorials which are a strength of this book will also work very well for people in industry that need to learn *P3e*. This book presents scheduling primarily from the general contractor's point of view, but most of the information can also be applied to the subcontractor, fabricator, owner, and construction manager.

ABOUT THE AUTHORS

David Marchman teaches estimating, scheduling, project management, and other construction courses in the Construction Engineering Technology program at The University of Southern Mississippi. His teaching career is supported by extensive experience in the construction industry (residential, commercial, and industrial), including seven years with Brown & Root. He continues to consult with numerous companies on scheduling. He has taught numerous construction-related seminars and workshops for such companies as W. G. Yates Construction, Mid-State Construction, Hill Brothers Construction, and the Associated Builders and Contractors of Mississippi.

Terry Anderson teaches construction organization, structures, hydraulics, project controls, and soils and foundations at The University of Southern

Mississippi. Prior to teaching he was a project superintendent for a national construction firm, a project manager for a large consulting engineering company, and an operator of his own consulting engineering practice for fifteen years. He is a registered professional engineer in six states and a registered land surveyor in two. He continues to provide consulting and forensic engineering services for private sector and governmental agencies as well as conducts frequent seminars for the major construction trade associations.

Acknowledgments

David Marchman—I thank my wife, Janet, for her patience and understanding. Thanks are also owed to our daughter Dee and our son Dane and his family (Holly, Forrest, and Sawyer) for their inspiration in the writing process.

Terry Anderson—I thank my wife, Jan, for her 25 years of support and encouragement. I also thank my children, Matt and Katie, for their understanding and love.

The authors and Delmar wish to thank the review panel for providing feedback and suggestions. The reviewers are:

Bill McMichael
Primavera Software Company

Dr. Sandra Weber
Del E. Webb School of Construction
Arizona State University
Tempe, AZ

Dr. Ihab Saad
Eastern Carolina University
Greenville, NC

Dr. Nancy Holland
Texas A & M University

Dr. Charles Patrick
Morehead State University
Morehead, KY

James Jenkins
Purdue University
W. Lafayette, IN

Section 1

Planning

Introduction to Scheduling

Objectives

Upon completion of this chapter, you should be able to:

- Explain the necessity for good scheduling
- Determine project scheduling needs
- Identify and name activities
- Identify different types of schedules

HISTORY OF SCHEDULING

Human Construction Capabilities

Construction scheduling is as old as human attempts to alter and control their environment. Prehistoric humans scheduled their lives around the cycles of nature that impacted their means of survival. Changes in seasons or long-term weather patterns, changes in herd movements, and changes in the availability of materials required a constant re-evaluation of the need to move and rebuild shelters and facilities.

As human's mastery of the environment progressed, so did the need for more detailed and accurate schedules. The Seven Wonders of the Ancient World are a tribute to humanity's ability to utilize tools, materials, and labor to alter and improve the environment we live in. Of those marvelous examples of construction, only the pyramids of Egypt still remain. Their construction required the efforts of thousands of skilled and manual workers and many years of planning and implementation. The techniques utilized by the Egyptians to construct these monuments are still not completely understood; however, it is widely assumed that the transportation of the huge stones necessary for the construction of the pyramids was heavily dependent on the cycles of the Nile River flood events. The commitment to build such enduring structures and the contributions of money, time, and other resources reflect the greatness of these societies.

An Example of Efficient Scheduling

Scheduling has always been an essential part of construction. Resources and time to complete projects are always limited. End users of every structure anxiously await its completion. Scheduling techniques, including cost management, have improved over the years to meet the demand for advanced structures. A scheduling feat that preceded the advent of both the computer and modern scheduling techniques was construction of the Empire State Building in New York City, the tallest building in the world at the time. Because the site was located in downtown Manhattan, there was already tremendous congestion. Without space on the site to store construction materials, the builders had to bring in the structural steel nightly to be used the next day. Compounding the site-related difficulties were labor problems. In addition, the owner wanted to use the building as soon as possible. This clearly was a project demanding good scheduling techniques. Remarkably, thanks to efficient scheduling, the builders managed to erect one floor every three days, on average.

Efficient scheduling not only provides better utilization of the owner's and contractor's resources but it can directly affect health, safety, and the quality of life. The Northridge earthquake of California in 1992 caused significant damage to the infrastructure of the region. Thanks in part to the use of constantly updated computerized scheduling, contractors were able to repair and restore essential facilities in a matter of weeks rather than years.

Progress in Scheduling

The size and nature of the projects that modern society constructs display the capabilities of people to dream, plan, and marshal their resources as they did in ancient times. Construction projects are getting larger and more complex all the time. Ancient construction projects primarily required huge amounts of labor resources, with the schedule being understood by only a few key individuals. With today's highly complex construction methods, it is necessary for everyone involved in the project to understand and use the constantly updated schedule. Recently, an ambitious tunnel was built underneath the English Channel. The size and complexity of this project required the use of the latest in technology, knowledge, and commitment. Whether of great or modest scope, no construction project can be accomplished without adequate planning and scheduling.

NECESSITY FOR GOOD SCHEDULING

Construction projects are becoming more and more complex and costly, requiring greater attention to the management of both time and resources.

A *resource* is "anything you need to execute the project for which you pay money to acquire." This includes information and time. Any type of construction (residential, commercial, industrial, small, or large) requires gathering resources such as specialized labor, materials (fabricated both on- and off-site), construction equipment, and site management. These resources are expended over the time that is needed to complete the project. Planning and controlling resources and pacing building to meet contractual deadlines and quality requirements are crucial to the success of any construction project. There is too much at stake to undertake a construction project without a detailed plan. Without a plan, there is no way to schedule the required resources, track actual progress, and decide on corrective action when unexpected events occur. Effective construction today requires close integration of planning, design, estimating, and scheduling to achieve project efficiency and profitability.

Time: The Owner's Viewpoint

The old adage "Time is money" is particularly true in construction. When a project is finished, the owner has use of the facility and starts to receive a return on investment from the use of the building to repay funds expended in the construction of the project. The *return on investment* is the money left over after the product or service is sold and the cost to produce the product has been deducted. The cost of constructing the facility used to produce a product or service is usually a large part of the owner's total investment in production.

Since the owner wants to produce a return as quickly as possible, another old adage applies: "Time is of the essence." This means that the project completion date is of concern to the owner, and any delay could result in damages that would be paid by the contractor to the owner for losses incurred. In a *lump-sum project*, the owner signs a contract with the constructor to produce a project for a set dollar value within a specified time. In a *negotiated cost-plus project*, the owner usually negotiates with the construction company to act as its agent to produce the project for a set fee in the shortest possible duration. In most construction contracts, whether lump-sum or negotiated, time is as important an element as cost.

Duration: The Contractor's Viewpoint

Typically, the contractor would also like to minimize the project's duration. Once the project is complete, the contractor's indirect costs stop. Indirect costs include temporary facilities and utilities, field supervision, and construction equipment and thus depend on project duration. The longer these resources are kept at the job site, the greater the cost. A crane might cost $1,000 per day to have at the job site. If the crane is only needed for 10 days rather than the 20 days scheduled, the project saves $10,000. The same work is accomplished but, through better scheduling or allocation of resources, money is saved. The accumulation of such savings can significantly reduce total project costs. (Of course, if a schedule is shortened by means of overtime, overall costs may not be reduced at all!)

A Service Industry

The construction industry is a service industry. The contractor is selling his or her firm's management ability to put the owner's project in place. For a traditional, lump-sum project, the owner furnishes the money, the architect and engineer furnish the contract documents, and the contractor furnishes the construction expertise. Any time the construction management team saves in meeting a project schedule can be spent on the next project. Saving time thus increases the volume of work and profitability. The efficient construction company can produce more work per year with the same management staff. Time *is* money!

Project Control

Good project management means "project controls" rather than a project "out of control." All projects have two primary control documents: the estimate and the schedule. The *estimate* defines the scope of the project in terms of quantities of materials, work hours of labor, and hours of equipment. Since all these resources can be measured in dollars, the estimate also serves as the cost budget.

The *schedule* is produced to control the expenditure of another resource: time. Since time is money, the resources and the time frame over which they are expended are directly related, and the elements of the estimate and the schedule are interrelated. The information from one document feeds the other; especially with method-related changes the estimate computes resources needed over the specified time, while the schedule defines time needed given the specified resource. These documents are the vehicle the project team uses to build the project on paper and in their minds before the actual construction begins. This is the essence of planning.

Unexpected Events.   Unexpected events are unwelcome at the construction job site. Estimating and, to an even greater extent, scheduling, provide the project organization with an efficient mechanism for anticipating and solving problems during the planning process before resources are expended—rather than on the job site during the construction phase. The process of planning is intended to make the construction process run smoothly, minimize surprises, and prevent the expenditure of costly resources with poor results. Effective project control combines both the estimate and the schedule to achieve a clear understanding of the status of the project at any given time.

Communication.   Another key to avoiding unwelcome and costly surprises is communication. Only the very smallest of construction projects is completed by a single individual. Most construction projects require the services of many different experts, various contractual relationships, and numerous functions. The project manager needs to provide and receive certain information at every stage. The same is true for superintendents, foremen, subcontractors, owners, and supervisors of the equipment and personnel departments. For a project to run smoothly, all parties need to be "on the same page at the same time." This requires continual communication so that all participants know what the plan is and what their responsibilities, requirements, and expectations are. Keeping everyone informed is one of the critical functions of the schedulers.

An effective schedule needs to reflect the status of the project at any given time and must be as clear as possible. A schedule with too much detail is overly complicated and is usually ignored. A schedule with too little detail is meaningless as a management tool and is also ignored. Finding the right balance of detail and essential information is a key role of the effective scheduler.

PROJECT MANAGEMENT

The purpose of project management is to achieve the project goals and objectives through the planned expenditure of resources that meet the project's quality, cost, time, and safety requirements. A constructor is in business to be successful and make money. This is accomplished through estimates and schedules that are fair, honest, and reasonable. Financial risks can be controlled and minimized to produce a successful project.

Quality

Good management assures that a project attain the level of quality as defined in the contract documents. Not only do quality and pride in workmanship go hand in hand and improve worker pride and project outcome, but they also lead to profitability through repeat business, referrals, and negotiated works.

Quality does not mean perfection. Quality means meeting the minimum requirements of the project as listed in the specifications. The contractor needs to remember that the owner is paying for, and expects, a certain level of quality. Anything above that level has a direct impact on the profitability of the project for the contractor.

Cost

The ability to estimate and then to complete a project within budget goes to the very heart of good construction project management. This is true whether the estimate is lump-sum or negotiated. The owner looks to the contractor for the efficient expenditure of resources to get the most results for his or her dollar. Construction is an industry fraught with incredible risks. Even seemingly small problems on the project can quickly turn into situations costing thousands of dollars if not handled quickly and correctly.

Time

Time is money! Time is of the essence! Time is usually just as critical to the contract and the owner's needs as are quality and cost constraints. The ability to determine a schedule and then to complete the project within the time frame also goes to the very essence of being a construction project manager. A properly formulated and executed schedule can help both the owner and the contractor to recognize and deal with disruptions far in advance of a pending catastrophic delay. Many times an owner chooses a contractor based upon his or her ability to marshal forces to complete a project in a timely manner, rather than basing the decision solely on the lowest cost. Getting the project in service and generating a return on investment as soon as possible is critically important to the owner. Tax or market considerations may concern the owner just as much as cost and quality.

Safety

Another integral component of good project management is safety. Accidents can be extremely costly, not only in their direct costs, but also because of their influence on the Workers' Compensation EMR (Experience Modification Rate). The rating of the construction company's safety record affects the amount the company must pay for insurance coverage for employees. Accidents also affect worker productivity, morale, and other direct and indirect costs. Good safety practices are an indicator of good project management.

SCHEDULING PHASES FOR DESIGN-BID-BUILD CONSTRUCTION

Design-Bid-Build

In *design-bid-build* construction, processes are consecutive rather than concurrent. There are breaks between the design, bid, and construction functions. The owner first formulates an idea for a potential project. After determining the feasibility of the project, the owner contracts with the architect and engineer to produce the project contract documents that define the project scope. Next, bids for construction services are accepted from contractors. The owner/architect selects the successful contractor's bid and awards a contract, and construction commences. Thus, the contractor knows the scope of the work, and the owner knows the price and duration of the work before the contract is awarded. For the constructor, scheduling lump-sum, linear projects deals primarily with scheduling field construction activities. The preconstruction activities of architecture/engineering design and the postconstruction activities of maintenance/operation are not part of the contractor's responsibility and therefore are not scheduled.

Field Construction

In controlling the field construction, the contractor is concerned with controlling time, resources, labor, materials, subcontractors, vendors, equipment, and money. It is critical to control such influences to the schedule as:

- The relationship of the activities to each other (which activities precede which, and which can occur at the same time)
- The size of crews
- The availability of labor
- Construction methods
- The cost of resources
- The types of construction equipment
- Work schedule (number of hours per day, shifts, weekends, holidays)
- Shop drawing review and approval
- Material deliveries
- Inspections
- Payment schedules

All these factors must be efficiently organized, sequenced, and controlled to optimize and maximize efficiencies.

Timing

On a lump-sum project, detailed scheduling typically begins once the contractor has signed the contract with the owner to construct the project. Although some scheduling takes place during the estimating/bidding phase, this is usually general in nature, for example, determining how long a superintendent will be needed on the project site.

Since producing the detailed schedule can cost a substantial amount, it is not begun until the contract is awarded, and it is completed as soon as possible. There are three primary reasons for preparing the schedule immediately. First, the contractor wants the schedule to control field operations. Second, the planning and scheduling process is a tool to help point out and solve problems before they arise in the field. Third, the owner/contractor contract may require the presentation of a project schedule to the owner/architect for use in monitoring the project.

The scheduling process used in setting up project controls for lump-sum, linear construction involves four phases: planning, scheduling, monitoring, and controlling.

Planning

Planning and scheduling efforts are done in conceptual project phases by the owner, the architect/engineer, or the construction manager to determine the expected project duration inserted in the contract. But, in a design-bid-build, planning for the contractor takes place during the stage prior to construction where the following must take place: decision making about the execution of the project; information gathering about the work to be accomplished; and identifying and defining activities. Good scheduling requires creativity and flexibility in identifying and defining activities and their interrelationships.

Decision Making.   Prior to starting construction, the management team must plan the execution of the project. Planning is a form of decision-making since it involves choosing among alternative courses of action. The effective project plan is always preceded by a carefully formulated preplan. The preplan gives project participants an opportunity to consider how the project fits into the company's overall operational strategy.

Information Gathering.   Planning entails defining the work to be accomplished. Communication is required to gather information from many persons and places. For example, questions might include: When and from what source is the right crane available? Are there enough skilled masons available to complete the brickwork on time? What will be the impact on the project schedule of long lead-time items, such as an elevator for the project that will require special fabrication at a shop where there is a backlog?

Identifying/Defining Activities.   Proper scheduling requires a thorough knowledge of construction methods as well as the ability to visualize individual work elements and establish their mutual interdependencies. The team preparing the schedule must create a rough diagram that identifies and defines activities and their relationships to other activities. Essentially, the project is built on paper with activities and interrelationships as the building blocks. The entire project is first constructed in the minds of the scheduling team, before being put on paper. The relationships among activities, building methods, problem solving, and communications that define the plan are generated. Activity relationships are defined using information derived from the contractor's management team, the accounting department, the equipment department, the plans and specifications, a visit to the site, the nature of the work, the owner, the banker, the subcontractors, the governmental agencies involved and their requirements, and the suppliers.

Creativity.   Good planning requires being creative and not being bound by preconceived notions. Just because a company has always tackled a certain construction sequence or segment of work a particular way, it need not assume that this is the only or best way. By breaking away from established paradigms, models, or false constraints, planners can incorporate improved methods into the schedule. One way to incorporate new ideas and methods is brainstorming with key participants to solve a specific problem. For example, perhaps the question is "What is the best way to pour the elevated concrete column on a specific project?" Participants in the brainstorming session each state the first method or solution to the problem that comes to mind, no matter how far-fetched or nontraditional the approach. This and other methods to expand and change the way team members look at the project are very useful in the planning process.

Flexibility.   Good planning also requires flexibility. It is the natural tendency of first-time schedulers to build a long chain of activities, one after another, with no branches, and with only one event happening at a time. The most efficient way to shorten project duration is to have as many activities going on concurrently as possible without their hindering each other's progress. Instead of single crews working consecutively at the job site, there could be five crews working concurrently but not interfering with each other's progress. In Figure 1–1, the activity Rough Framing Walls comes after Pour Slab. They are scheduled consecutively because the first must be complete before the other can begin. The activities Rough Plbg, Ext Fin Carp, Rough Elect, Rough HVAC, and Inst Wall

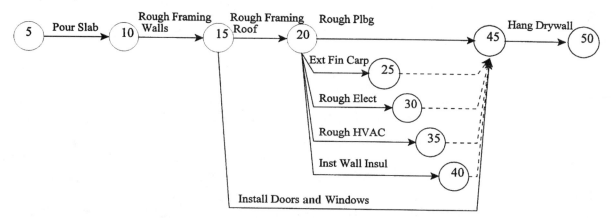

Figure 1–1   Activity on Arrow Diagram—Example of Consecutive and Concurrent Activities

Insul are scheduled concurrently, since they can be going on at the same time.

Interrelationships.   During the planning stage, durations are not applied to the schedule. The key development at this stage is naming (identifying/defining) activities and establishing relationships to the other activities. In establishing activity relationships, resource requirements must be considered. For example, an activity might be either cast-in-place concrete or precast concrete. Thus, some resource decision has already been made, but the activities are usually only defined along with their associated interrelationships and constraints. This is the phase of the scheduling process with the most potential for the creation and development of new approaches, systems, or methods for putting the work together. When the project has been defined on paper in the form of the rough schedule, the planning stage gives way to the scheduling phase.

Scheduling

Durations of Activities.   The second phase of the overall scheduling process involves filling in more precise estimates of the time and other resources in the rough schedule produced in the planning phase. The other resources might include labor, equipment needs, or division of responsibility (for example, crew versus subcontractor). By calculating the estimated duration of each activity, the scheduler can calculate the project duration. If the first pass yields a project duration that is too long, the original logic constraints may have to be adjusted (for example, by increasing crew size to complete an activity sooner, by modifying sequences of activities, or by making activities concurrent). The schedule is continually fine-tuned until it becomes satisfactory to all parties; it is then accepted as the project schedule.

Evolution of the Schedule.   Development of the schedule from the rough stage to the project schedule is an evolutionary process that requires communication and the approval of all parties involved—the contractor's management team, subcontractors, major suppliers, and, of course, the owner. The contractor must receive input from these parties, and the contractor wants each of the parties to accept or "buy into" the schedule. This process makes it "our schedule" instead of "your schedule." Typically the contract documents make the contractor the keeper of the schedule for the construction project. It is the contractor's responsibility to resolve scheduling disputes between the different subcontractors and the contractor's forces. Thus the process includes soliciting input from the different parties, drafting a rough schedule, circulating it for feedback, making modifications, resolving disputes, and reaching agreements. This process produces the project schedule that all parties can accept.

A properly prepared construction schedule always proceeds from a small number of general activities to a more detailed number of specific activities. For example, in the early stages of schedule preparation a single activity may be listed as "concrete slabs." As the schedule is developed, this activity may be further broken down into "erection of forms," "placement of reinforcing steel," "concrete placement," "finishing," "curing," and "stripping." A good scheduler will seek to achieve a balance between a schedule that is so general that it does not yield adequate information and one that is so specific that it hides the truly important planning factors.

Monitoring the Schedule

Actual Progress. The third phase in the scheduling process is the monitoring phase. The *data date* is the date on which a schedule is updated with current information. Typically, at fixed intervals throughout the project (usually at the end of each month) progress is determined as of that date. This is the cutoff date for comparing the actual project progress to the planned progress. Determining actual activity progress can be done in several different ways. The following describes the expenditure of resources. Monitoring durations/resources involves determining the physical progress in the field and inputting the progress of each activity. First, it is necessary to establish a database of the actual expenditure of resources for each activity. This information is compared to planned expenditure to determine the percent complete for each activity. The progress of each individual activity is established using the information from the database. Next, the schedule is recalculated with the updated information to determine if each activity is ahead of or behind schedule when compared with the plan (baseline) and if the overall project is ahead of or behind schedule. This process is called *updating the schedule*.

Controlling the Schedule

Schedule Changes. The fourth and last phase in the scheduling process is the controlling phase. Controlling usually involves documenting and communicating changes to the plan and schedule. As projects develop, the sequence of activities originally planned may change. The reason may be updating of the schedule, changes in scope, material delays, lower or higher productivity, or some other factors. An example of a change may be the decision to use a different erection method or system, which essentially alters the schedule. This new, revised schedule then becomes the current project schedule. If the project is behind schedule, the contractor may have to "crash" or accelerate the schedule to make up for lost time by adding shifts, having laborers work overtime, or adding craftspeople to a crew. Keeping the schedule relevant and useful requires redrawing the plan to incorporate such changes in the relationships. The new, revised schedule will be better, since it is based on more current information.

Schedule Progress. Usually schedule progress is the basis of monthly meetings with the owner/architect to determine the contractor's compliance with the contract and invoice for payment. The owner is presented with an updated schedule showing the data date, progress during the last 30 days, and the forecast for the next 60 days. The schedule is a critical document for determining whether the project is progressing according to the original plan. By using the current (updated) schedule, along with the target (original budgeted) schedule, it is easy to spot which activities are in trouble and whether the project itself is on schedule or in trouble. This concept is known as *management by exception*. Management and usually the owner want to know which parts of the schedule are in trouble so they can determine which activities to spend their time on and which activities are the most likely candidates for making up lost time. It does not do any good to put extra resources on an activity that is not critical to the completion of the project schedule. Controlling is the process of constantly modifying the schedule to make sure it is current with the latest plan for how the project will be constructed, and in case of deviation from the original schedule, suggesting and implementing a corrective action.

Documentation. When changes are made, the construction team must document or record the changes for historical information and project backup. The historical information is used as a database for reference for future projects. The project backup is necessary for use in possible legal claims or settlement of project disputes. The importance of adequate documentation cannot be overemphasized. Many cases of construction litigation have been resolved in favor of the party having the most legitimate and thorough documentation.

SCHEDULING PHASES FOR NEGOTIATED, FAST-TRACK CONSTRUCTION

The previous section of this chapter dealt with sequential or design-bid-build construction. The scheduling for the usual negotiated/fast-track construction requires a different mind-set. The negotiated/fast-track construction is discussed in this section.

Integration of Activities

Fast-track or phased construction involves the integration of detailed design and construction. On larger projects, significant time can be saved by overlapping or concurrently designing and building. The primary disadvantage is that, since the design is not complete, the full scope of the work is not known before field work commences. These projects are, therefore, negotiated contracts, with the owner accepting a greater portion of the risk for project cost increases.

Team Approach to Construction

Another advantage of fast-track construction is the team approach to the design/construction process. Teamwork among the owner, designer, and constructor contrasts with the confrontational relationship that lump-sum construction typically leads to. The constructor is typically paid a fee for services and is part of a team. Under the lump-sum arrangement, the constructor is paid according to bid and receives money saved. This leads to the construction firm looking out for its own best interest and not that of the owner. Under the team concept, if the contractor's fee is fixed, money saved returns to the owner, or can be shared between owner and contractor, and the constructor is looking out for the best interest of the owner.

Team Approach to Design

Another advantage of fast-track construction is constructor involvment in the design process enhancements. Significant cost savings can be derived through "constructibility"—designing for ease and economy of construction. Another way to save is by value engineering. *Value engineering* is a systematic approach to evaluating a number of cost alternatives to choose the best one for the particular project. An added advantage of value engineering is that, in many construction contracts, money saved on the project is shared with the contractor.

A good rule of thumb is that 80 to 90% of a project's cost should be fixed by the time conceptual design (the sketching phase) is completed. The project cost is fixed to a large extent before the first detailed drawing is ever completed. This is because when certain project parameters are defined, the scope and therefore the cost of the project become fixed. Once the function, location, size, vertical or horizontal orientation, type of construction, and type of environmental controls are fixed, the project cost is essentially determined. Since the constructor's knowledge of cost and how the project goes together is likely to be greater than that of other members of the team, the constructor's input during the conceptual design phase is invaluable in controlling project cost and schedule.

Comparison of Lump-Sum and Negotiated Projects

The four phases of scheduling (planning, scheduling, monitoring, and controlling) are the same for negotiated and lump-sum projects. The primary difference is that the scheduling effort in the lump-sum project focuses on the field construction effort, whereas a negotiated schedule integrates the conceptual design, detailed design, project management, procurement, expediting, documentation, field construction, and possibly maintenance/operations. The entire project is looked at as a whole, with break-out schedules for each of the areas such as design, construction, or project management.

SCHEDULING LEVELS

During the control phase of the schedule process, most contractors consider the project schedule established at the beginning of the project as the general or starting point. Good scheduling requires more detailed preparation as a particular activity gets closer to actual installation.

It is good practice to prepare a written *10-day (2-week) look ahead* or a preplanning sheet (short-interval schedule) of all upcoming activities. Depending on the size and complexity of the project, the preplanning sheets are organized by some combination of the project engineer, superintendent, and foreman. This short-term schedule is usually a field function and is used for communication with the crews. Good preplanning techniques require that all necessary sketches or shop drawings be complete, the plan be in writing, and materials be available for the work to commence. The preplanning process is essentially a problem-solving exercise to save time and money by averting problems that might arise in the field. Preplans should be in writing to reduce mistakes and rework. A major cause of rework in the construction industry is the heavy reliance on oral communication rather than written communication. Workers try to put in place what they think their supervisors want; however, since they do not always understand exactly what the supervisors want, they sometimes complete the wrong work. Precise written plans can prevent such miscommunication. Also, requiring that instructions be reduced to writing requires the supervisor to give more careful thought to his or her directives.

ACTIVITIES

One of the first steps in putting any schedule together is identifying the activities or tasks that must be completed to attain the project goals and objectives of the project team. Since this book is based on Primavera Systems, Inc.'s project management/scheduling software package, *Primavera Project Planner for the Enterprise (P3e)*, activity definitions and other relationships used in this book reflect the use of this software. This book is based on *Primavera Project Planner for the Enterprise*, Release 2.0.

Activity Defined

Construction projects are made up of a number of individual *activities* that must be accomplished in order to complete the project. The primary activity type is the *task-dependent activity*, which requires time and usually resources to complete. Task-dependent activities have five specific characteristics. They must:

- Consume time
- Consume resources
- Have a definable start and finish
- Be assignable
- Be measurable

Activity 1000, Clear Site, in Figure 1–2 is an example of a task activity. Tools used to help define activities are:

- The scope of work
- The estimate
- Historical information
- Work breakdown structure (WBS)
- Experience

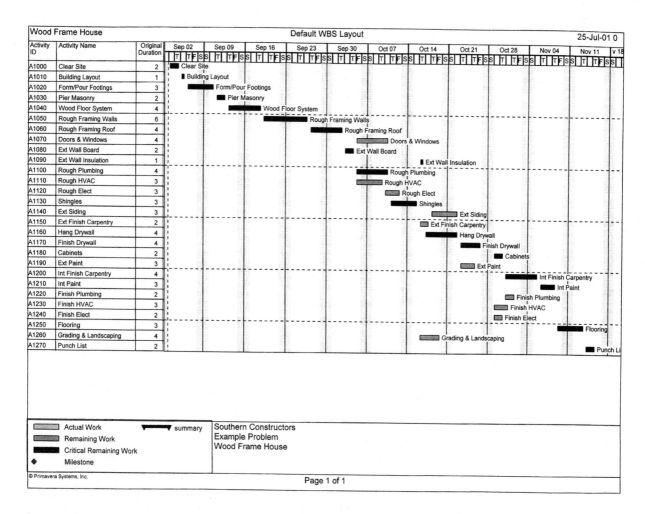

Figure 1–2 Gantt Chart—Wood-Framed House

The *P3e* system also uses the *milestone activity*, which is a major event, phase, or any other important point in the project.

Time Consumed.   The task activity breaks the schedule down into more easily estimated smaller components. A task activity may consume weeks, days, or hours. Duration is a function of the scope of work for the activity and resources assigned to accomplish the work.

Resources Consumed.   Usually resources must be expended for a task activity. The assigned resources should be scheduled according to the activity's base calendar. Labor is expended to install material and equipment resources to complete the finished structure. The scope of work, as defined by the estimate of labor, materials, and equipment, determines the duration of the activity.

Definable Start and Finish.   Task activities consume time and resources and are tied to related activities by relationships. These relationships determine which activities must be complete before the activity in question can begin. The scheduler determines duration of the activity from the estimated quantities of materials to be placed and the size of the crew to place the materials. By knowing the relationships between the activities and their durations, the scheduler can determine the planned start and finish dates of the activity. Therefore, each activity is definable in terms of its planned start time, duration, and planned finish.

Assignability.   Determining responsibility for each activity is critical to any scheduling effort, since any construction project involves bringing together many craftspeople, subcontractors, suppliers, and others to attain project completion. Task activities should be defined so that the responsibility for activity completion is clear and assignable to a single party. If the activity is to be completed by the contractor's own forces, it should be defined by the crew to identify the proper superintendent or foreman. If the activity is to be completed by a subcontractor or vendor, responsibility should be assigned to that party, so that as the project progresses, communication can be expedited using the schedule as the baseline indicator of the performance. The plan is something to measure against to determine performance.

Measurability.   The duration and resources assigned to the activity must be measurable to determine whether the budget for the activity duration was met. How many days were actually spent? What was the actual physical progress in terms of days? How many resources were actually expended? What was the actual physical progress in terms of resources? The duration and resources are measured as of a particular date, and an evaluation is made about how the project is going in terms of the original plan. Is the project ahead of or behind the original schedule with respect to time? Is the project ahead of or behind the original schedule in terms of the expenditure of resources? Answering these questions is essential for monitoring and controlling the project. Since the activities are assigned, the party responsible for controlling the duration and the expenditure of resources is identified by activity, and communications to resolve project control problems are enhanced.

Activity Codes/Categories

It is usually not enough to identify only those activities that fit into the overall logic of the project. It is also important to assign responsibility for the completion of an activity. Good management requires defining authority and responsibility by activity so that communications about the activity and the management chain of command are clear. This is clearly a choice for using activity data and is not mandatory. When responsibility is assigned by activity, *P3e* can provide specialized information to give the responsible parties copies of only their activities so that they will not have to dig through mountains of data to find the information they require.

P3e has the built-in capability to sort activities defined by project requirements. The usual sorting parameters of department, responsibility, phase, work breakdown structure, or other requirements can be defined in the system. In addition, establishing project activity categories/codes and assigning related activities gives the project management team the ability to sort the schedule by category. The party responsible for a specific portion of the schedule can concentrate on the activities for which he or she has control.

Engineering.   The engineering category of activities usually relates to construction permits, project layout and field control, documentation control (for example, tracking shop drawings), payment, and inspection. Often these activities or events fall into the category of a milestone activity, an event that enables other activities to continue or commence. For example, the visit by the off-site inspector enables the rest of the activities to begin. Even though the inspection had no cost or resources connected to it, the occurrence can be considered a milestone activity.

Mobilization.   The mobilization category of activities relates to moving onto the job site. Besides bringing in construction equipment and temporary materials, it also includes installation of temporary facilities and utilities. The majority of the mobilization effort occurs early in the life of the job; however, some work items may require mobilization at various later stages in the life of the project.

Procurement.   The procurement category of activities includes the identification, procurement, expediting, delivery, and control of bulk materials, fabricated materials, and permanent equipment to be used on the job site.

Construction with a Contractor's Own Forces.   Work in this category is performed by craftspeople who report directly to the contractor and not to a subcontractor. The contractor sorts the activities by the foreman responsible for carrying out the work. The labor resources expended are sorted by type, such as rough carpentry, finish carpentry, concrete finishing, etc.

Construction with Subcontractors.   It is critical to establish the subcontractor's responsibility for the activity in this category. It is a good practice to document the labor to be expended by the subcontractor's craft classification. The contractor is not responsible for the direct supervision of the craftspeople, but by comparing the expenditure of resources according to the subcontractor's original plan, the contractor can determine if the subcontractor is within budget. Comparing actual with anticipated expenditure of resources also enables the contractor to assess physical progress and evaluate payments.

Start-Up.   The start-up category of activities relates to testing, punchlist, and start-up of the facility. Start-up is the point where the owner takes possession of the facility to use for the intended purpose. There are many activities related to testing, accepting, and starting the facility, as well as activities related to the owner taking responsibility for insurance, utilities, security, and possession of the structure.

Demobilization.   The demobilization category of activities relates to moving off the job site. This includes the removal of temporary facilities, construction equipment, and temporary materials.

Activity Identification

The activity identification (activity ID) is the way the activity logic is identified to the computer. The activity ID is used by *P3e* (and most other scheduling software programs) to give the activity a short name or identifier that can be used for sort functions. In Figure 1–2, note that the Clear Site activity has an activity ID of 1000. An example of using this field as a sort field would be if an owner wanted to break a project into three phases (A, B, and C). All activities relating to phase A would start with the identifier A, such as A1000. All activities relating to phase B would start with the identifier B. This is a handy tool for quick activity identification.

P3e maintains a database of information about each activity, and the key or primary field used to sort the information is the activity ID field.

The Activity ID itself can be a convenient means to sort activities. Take care when developing the naming format to make the activity ID a sortable field. Be consistent in naming all project activities to make the sort possible.

Activity Detail

Activity detail defines the appropriate level of information breakdown needed to meet the project needs. Usually, a daily unit is appropriate, and units less than a day should be consolidated if possible. Other considerations to keep in mind besides the daily unit are:

- Who is going to use the schedule and what are their needs?
- Complexity: What level of communication is needed?
- Division of responsibility: Who is doing the work?
- What is the contractor's management philosophy?
- Will more or less information affect the usefulness of the schedule?
- Will short-term scheduling be appropriate for more detailed scheduling?

Activity Name

The activity name is a *P3e* field that is longer than the Activity ID field and is used to describe the activity. (See Figure 1–2.) An important communications tool, it must be clear, concise, and have the same meaning to all parties using the schedule. This includes the contractor's forces, subcontractors, owner, and the architect/engineer.

The activity name, 120 characters in *P3e*, must communicate the scope and location of the portion of the work that the activity encompasses. Because so much information is communicated, descriptions must be consistent in format, and abbreviations are commonly used. Abbreviations and procedures for naming activities should be consistent throughout the project, as this will make the schedule much easier to use. Whenever possible, use standard industry abbreviations, such as "Ftg" for footing and "Conc" for concrete.

Activity Relationships

The relationships among activities determine which other activities must come before, after, or can be going on at the same time as the activity being defined. Again, activity relationships are defined using information derived from the contractor's management team, the accounting department, the equipment department, the plans and specifications, a visit to the site, the nature of the work, the owner, the banker, the subcontractors, the governmental agencies involved and their requirements, and the suppliers. These relationships give the schedule its "logic" to enable the calculations to work. The activity relationships determine the interaction of the parties to the work and to each other.

GANTT CHARTS

The Gantt chart is a convenient and easy-to-read method of viewing the schedule in bar chart form. Henry L. Gantt popularized this graphical representation in the early 1900s. Simply put, the horizontal axis represents a time scale of the project and the vertical scale lists the activities necessary to put the project together. These activity descriptions can be as broad or narrow as the scheduler needs to adequately describe the project. Figure 1–2 is an example of a *P3e*-generated Gantt chart.

Advantages of the Gantt Chart

There are three primary advantages to using the Gantt chart. First, it is usually easy to read. It is apparent when the activities should take place. Anyone involved in the construction process—owner, architect, banker, bonding agent, contractor, subcontractor, or supplier—can interpret this simple document. A second advantage is that, because of its simplicity, it is a great communications tool. The third advantage is that it is easy to update the chart to show progress.

Disadvantages of the Gantt Chart

The primary disadvantage of the Gantt chart is that it does not show the interrelationships among the various activities. What happens if one of the activities is completed late? How is the rest of the project delayed? The impact of the delay can be evaluated by analysis of the Gantt chart, but since the relationship between the activities is not shown, the conclusions are open to debate. The logic of the interrelated activities may be very formalized, but it is not clearly and completely conveyed to the user of the Gantt chart. A great majority of the construction claims relating to schedules that are lost by contractors are lost because the contractor cannot prove the impact of schedule delays. A construction claim requires proof through documentation. For example, if the drywall subcontractor is late hanging the drywall, typically all interior work is delayed. This would have an impact on finishing the drywall, placing cabinets, interior finish carpentry, and the rest of the schedule, but because the Gantt chart does not show the direct impact on these activities, there is room for argument and the claim of the contractor is more difficult to prove. A better tool for proving impact cost or ripple damages is a PERT diagram, which is discussed in the next section.

Gantt Chart Format

The Gantt chart format (along with the PERT and time-scale logic diagrams) is one of *P3e's* primary hard-copy graphical print formats. Once the scheduler has input the activity information into *P3e,* any of the three print formats can be requested. The Gantt chart is a convenient vehicle for confirmation and dissemination of the information used for more complex formats. Figure 1–2 is an example of the *P3e* hard-copy print of a Gantt chart for typical residential construction.

PERT Diagrams

The graphics for logic diagrams within *P3e* are organized for activity-on-node diagrams. The nodes (rectangles in Figures 1–3a and 1–3b) are the activities, and they are connected by arrows that show relationships between and among activities. The nodes may also contain information about the activities. The activity-on-node diagrams originated as the Program Evaluation and Review Technique (PERT) diagrams. PERT diagrams were developed for projects where activity durations and scope of work could not be determined with great accuracy, such as for new types of projects that had never been built before. The full extent of the work—or the relationships between the activities—was not understood. This method was originally used by the Special Projects Office of the Navy Bureau of Ordnance in the late 1950s and early 1960s in the development of the Polaris missile.

The node or box is the activity, and the arrows connecting the boxes show the relationships between the activities. Compare Figure 1–3a and Figure 1–3b to Figure 1–1 to see the difference between the activity-on-node and activity-on-arrow diagramming. The activity-on-node diagram is *P3e's* PERT diagram. Figure 1–3a and Figure 1–3b are examples of a hard-copy print of *P3e's* PERT diagram.

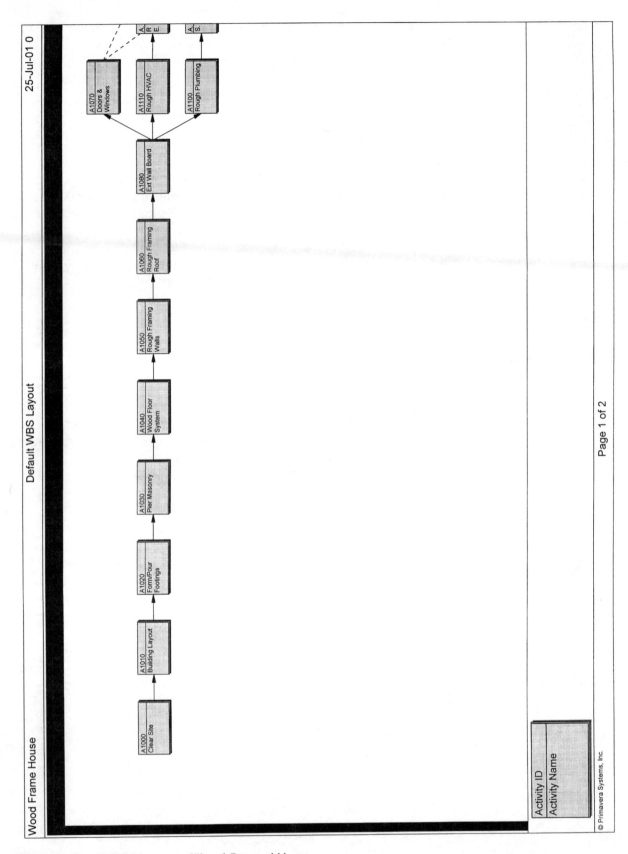

Figure 1-3a PERT Diagram—Wood-Framed House

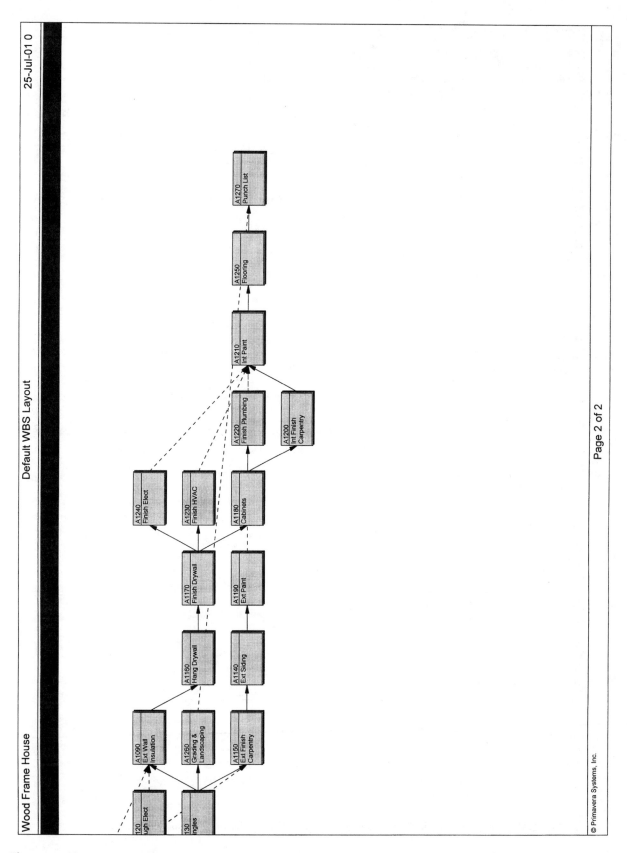

Figure 1–3b PERT Diagram—Wood-Framed House

ACTIVITY-ON-ARROW DIAGRAMS

P3e does not support arrow notation. The activity-on-node diagram is neater and more efficient from a graphical point of view. It is much easier to design graphics by putting information into a box, rather than associating it with a line of unknown length with which you are also trying to show logical relationships. Because of this, the activity-on-arrow diagrams are falling into disuse.

The activity-on-arrow diagramming method for scheduling was developed by the E. I. du Pont de Nemours Company in the late 1950s and was based on the Project Planning and Scheduling method (PPS). This form of scheduling was eventually called the Critical Path Method (CPM) (Figure 1–1).

TIMESCALED LOGIC DIAGRAMS

P3e's timescaled logic diagram (Figure 1–4) combines the advantages of the bar (Gantt) chart and the pure logic diagramming methods (PERT). Like the bar chart, it shows the activities' relationship to time, either in work days or calendar days. It also shows the relationships among activities.

The advantages of the timescaled logic diagrams are that they are easy to understand, like the bar chart, and they define the logic. For updating and documentation purposes, this is a tremendous advantage. The disadvantage, however, is that with increasing the size of the project, and therefore the number of activities, it becomes very hard to read and follow the logic. Figure 1–4 is an example of a hard-copy print of *P3e's* timescaled logic diagram.

TABULAR REPORTS

Sometimes a table or tabular report is the easiest way to communicate or update information. It is also easier to catalog in the project's historical database. Figure 1–5 is an example of a tabular report that lists logic information associated with each activity.

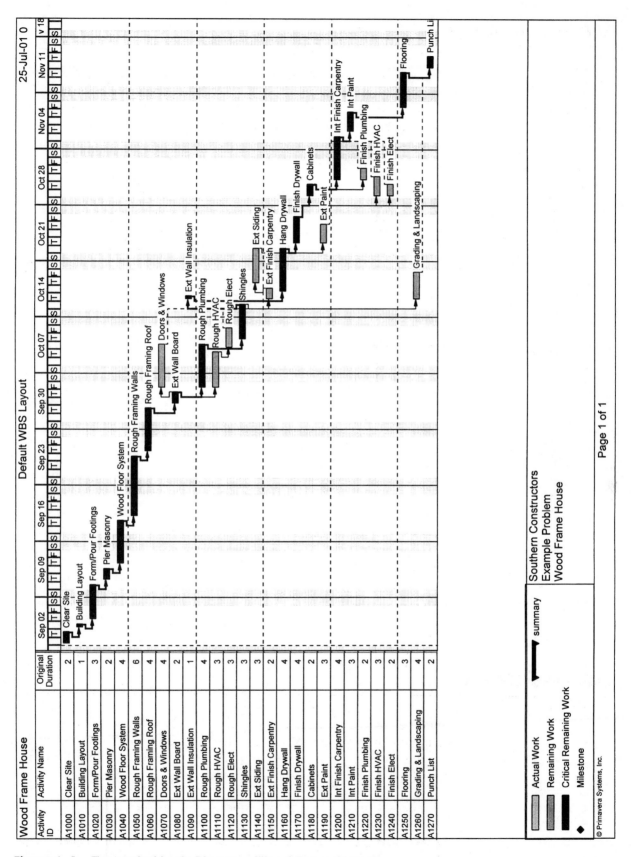

Figure 1–4 Timescaled Logic Diagram—Wood-Framed House

Wood Frame House
Report Date 25-Jul-01 08

Group: Student Constructors

LG-01 Logic Report, By Project

Project Start 03-Sep-01
Project Finish 15-Nov-01
Data Date 03-Sep-01

Activity ID	Activity Name	Early Start	Early Finish	Late Start	Late Finish	Total Float	Predecessors	Successors
Wood Frame House								
A1000	Clear Site	03-Sep-01 08	04-Sep-01 17	03-Sep-01 08	04-Sep-01 17	0d		A1010
A1010	Building Layout	05-Sep-01 08	05-Sep-01 17	05-Sep-01 08	05-Sep-01 17	0d	A1000	A1020
A1020	Form/Pour Footings	06-Sep-01 08	10-Sep-01 17	06-Sep-01 08	10-Sep-01 17	0d	A1010	A1030
A1030	Pier Masonry	11-Sep-01 08	12-Sep-01 17	11-Sep-01 08	12-Sep-01 17	0d	A1020	A1040
A1040	Wood Floor System	13-Sep-01 08	18-Sep-01 17	13-Sep-01 08	18-Sep-01 17	0d	A1030	A1050
A1050	Rough Framing Walls	19-Sep-01 08	26-Sep-01 17	19-Sep-01 08	26-Sep-01 17	0d	A1040	A1060
A1060	Rough Framing Roof	27-Sep-01 08	02-Oct-01 17	27-Sep-01 08	02-Oct-01 17	0d	A1050	A1080
A1070	Doors & Windows	05-Oct-01 08	10-Oct-01 17	10-Oct-01 08	15-Oct-01 17	3d	A1080	A1090, A1150
A1080	Ext Wall Board	03-Oct-01 08	04-Oct-01 17	03-Oct-01 08	04-Oct-01 17	0d	A1060	A1070, A1100, A1110
A1090	Ext Wall Insulation	16-Oct-01 08	16-Oct-01 17	16-Oct-01 08	16-Oct-01 17	0d	A1070, A1120, A1130	A1160
A1100	Rough Plumbing	05-Oct-01 08	10-Oct-01 17	05-Oct-01 08	10-Oct-01 17	0d	A1080	A1130
A1110	Rough HVAC	05-Oct-01 08	09-Oct-01 17	08-Oct-01 08	10-Oct-01 17	1d	A1080	A1120
A1120	Rough Elect	10-Oct-01 08	12-Oct-01 17	11-Oct-01 08	15-Oct-01 17	1d	A1110	A1090
A1130	Shingles	11-Oct-01 08	15-Oct-01 17	11-Oct-01 08	15-Oct-01 17	0d	A1100	A1090, A1150, A1260
A1140	Ext Siding	18-Oct-01 08	22-Oct-01 17	29-Oct-01 08	31-Oct-01 17	7d	A1150	A1190
A1150	Ext Finish Carpentry	16-Oct-01 08	17-Oct-01 17	25-Oct-01 08	26-Oct-01 17	7d	A1070, A1130	A1140
A1160	Hang Drywall	17-Oct-01 08	22-Oct-01 17	17-Oct-01 08	22-Oct-01 17	0d	A1090	A1170
A1170	Finish Drywall	23-Oct-01 08	26-Oct-01 17	23-Oct-01 08	26-Oct-01 17	0d	A1160	A1180, A1230, A1240
A1180	Cabinets	29-Oct-01 08	30-Oct-01 17	29-Oct-01 08	30-Oct-01 17	0d	A1170	A1200, A1220

Page 1 of 2

(c) Primavera Systems, Inc.

Figure 1–5   Tabular Report—Wood-Framed House

EXAMPLE PROBLEM: GETTING READY FOR WORK

Table 1–1 is a list of activities for getting ready for work. Figures 1–6 to 1–8 are completed samples of a bar chart, a pure logic diagram (PERT), and a timescaled logic diagram using the list of activities from Table 1–1.

	Activity Name	Duration (Minutes)		Activity Name	Duration (Minutes)
1.	Turn Off Alarm	1	12.	Place Underwear	1
2.	Get Out of Bed	1	13.	Place Shoes	1
3.	Remove Pajamas	1	14.	Place Shirt	1
4.	Brush Teeth	2	15.	Place Pants	1
5.	Take Shower	5	16.	Place Tie	1
6.	Wash Hair	2	17.	Take Vitamins	1
7.	Make Coffee	2	18.	Fix Cereal	1
8.	Perk Coffee	5	19.	Eat Cereal	3
9.	Drink Coffee	10	20.	Make Bed	2
10.	Dry Body/Hair	3	21.	Leave for Work	1
11.	Comb Hair	1			

Table 1–1 Activity List—Getting Ready for Work

Activities	1	2	3	4	5	6	7	8	9	10	11	12	13	14	15	16	17	18	19	20	21	22	23	24	25	26
Turn Off Alarm	×																									
Get Out of Bed		×																								
Remove Pajamas			×																							
Brush Teeth													×													
Take Shower					×	×	×	×	×																	
Wash Hair								×	×																	
Make Coffee			×	×																						
Perk Coffee				×	×	×	×	×																		
Drink Coffee										×	×	×	×	×	×	×	×	×	×							
Dry Body/Hair										×	×	×														
Comb Hair													×													
Place Underwear														×												
Place Shoes																	×									
Place Shirt																×										
Place Pants															×											
Place Tie																		×								
Take Vitamins													×													
Fix Cereal																				×						
Eat Cereal																						×	×	×		
Make Bed																								×	×	
Leave for Work																										×

Figure 1–6 Example Problem—Bar Chart Format

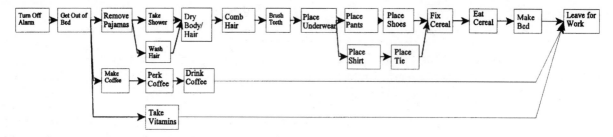

Figure 1–7   Example Problem—Pure Logic Diagram Format

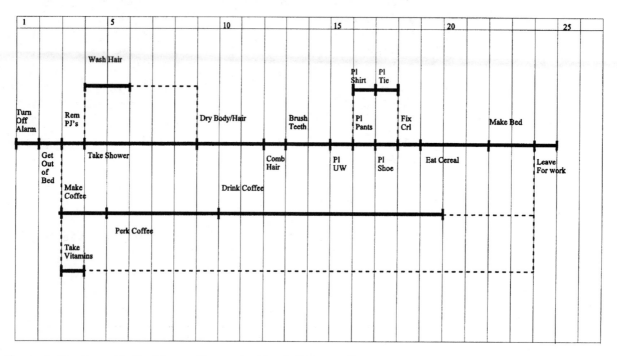

Figure 1–8   Example Problem—Timescaled Logic Diagram Format

EXERCISES

1. Building a Shed

The objective of this exercise is to organize and sequence activities. Using the tabular list of activities from Table 1–2, produce a bar chart and a pure logic diagram. Refer to the previous example problem for the format of each of these schedule types. You are to assume the precedence relationships.

2. Purchasing a New Automobile

The objective of this exercise is to organize and sequence activities. Using the tabular list of activities from Table 1–3, produce a bar chart and a pure logic diagram. Refer to the example problem for the format of each of these schedule types. You are to assume the precedence relationships.

	Activity Name	Duration (Days)
1.	Clear Site	1
2.	Remove Topsoil	1
3.	Form Slab	2
4.	Place Rebar/Embeds	1
5.	Pour Slab	1
6.	Prefab Wood Walls	2
7.	Erect Wood Walls	1
8.	Install Siding	2
9.	Place Trusses	1
10.	Place Roof Sheathing	2
11.	Place Interior Paneling	1
12.	Ext Trim Carpentry	2
13.	Install Overhead Door	1
14.	Rough Electrical	1
15.	Finish Electrical	1
16.	Place Shingles	2
17.	Install Finish Carpentry	1
18.	Place Topsoil/Grade	1
19.	Landscape	1

Table 1-2   Activity List—Building a Shed

	Activity Name	Duration (Days)
1.	Decision - Type Car	5
2.	10 Models - Make List	1
3.	10 Models - Obtain Consumer Ratings	1
4.	10 Models - Obtain Pricing Publication	1
5.	10 Models - Talk to Vehicle Owners	1
6.	3 Models - Decision	2
7.	3 Models - Test Drive	1
8.	3 Models - Obtain Information	1
9.	3 Models - Compare Lease/Purchase Options	1
10.	1 Model - Decision	2
11.	1 Model - Negotiate Purchase Contract	1
12.	3 Institutions - Compare Financing	1
13.	Decide on Institution	1
14.	Obtain Financing	1
15.	Money for Down Payment	1
16.	Obtain Insurance	1
17.	Obtain Tag	1
18.	Drive Away with Purchased Vehicle	1

Table 1-3   Activity List—Purchasing a New Automobile

2

Rough Diagram Preparation—An Overview

Objectives

Upon completion of this chapter, you should be able to:

- Enumerate project phases
- Decide which schedule format to use
- Determine schedule information needed
- Run schedule meetings
- Estimate activity durations

TYPES OF ACTIVITY RELATIONSHIPS

Logic refers to the entire network of relationships between and among activities. This network controls the scheduling of activities. When you prepare the schedule, you establish relationships between activities. These relationships describe which activities depend on others. The relationships between activities are defined as predecessor, successor, or concurrent relationships. *Concurrent activities* are logically independent of one another and can be performed at the same time. A *predecessor activity* is one that must be completed before a given activity can be started. A *successor activity* is one that cannot start until a given activity is completed. Relationships between a particular activity and its predecessor and successor activities can vary. The options, as shown in Figure 2–1, are:

- Finish to Start (FS)
- Start to Start (SS)
- Finish to Finish (FF)
- Start to Finish (SF)

The FS relationship means that the predecessor activity must finish before the successor activity can start. The SS relationship means that both activities can start at the same time. The FF relationship means that both activities can finish at the same time. The SF relationship means that the predecessor activity must start before the successor activity can finish. We will use Figure 2–2 and the information in Table 2–1 as an example throughout the chapter.

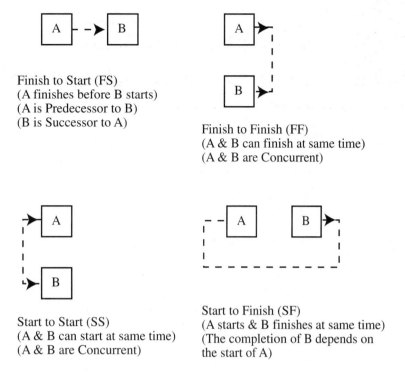

Figure 2–1   Activity Relationship Types

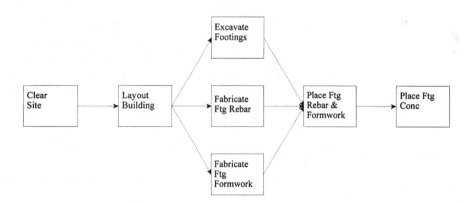

Figure 2–2   Logic Example

Clear Site	Predecessor to	Layout Building
Layout Building	Predecessor to	Excavate Footings Fabricate Ftg Rebar Fabricate Ftg Formwork
Excavate Footings Fabricate Ftg Rebar Fabricate Ftg Formwork	Predecessor to	Place Ftg Rebar
Fabricate Ftg Rebar	Predecessor to	Place Ftg Conc

Table 2–1   Logic Statements

ROUGH DIAGRAM

The four steps of project control are planning, scheduling, monitoring, and controlling. The very first step of the planning phase is the preparation of a rough diagram. A rough diagram builds the project on paper. It defines the activities and the relationships (logic) between them and the time to complete them. Once accepted, developed, and refined, the rough diagram becomes the project plan. Durations and resources are then applied in scheduling. During the rough diagram phase of the planning, the approach to placing the project takes shape. The planner should always remember that there should be a balance between the intricacy of the project and the number of activities scheduled. Too few activities in the rough diagram results in a schedule insufficient in detail to be worthwhile. Likewise, too many activities result in confusion and needless detail. In either case the schedule is likely to be ignored by those it is intended to benefit.

Construction decisions concern:

- Construction methods
- Flow of materials
- Prefab versus on-site assembly of materials
- Types of construction equipment
- Crew size and balance
- Productivity expectations
- Subcontractor definition

Management decisions include:

- Work breakdown structure (WBS)
- Division of responsibility
- Division of authority
- Software to be used
- Level and distribution of reports
- Interface with other functions (payroll, accounting, etc.)

APPROACHES TO ROUGH DIAGRAMMING

The four most common approaches to handling the rough diagramming meeting are using a tape recorder, making a list of activities, drawing the actual diagram, and using a software integrator.

Tape Recorder

Some schedulers prefer to record the initial meeting of key project participants on audiocassette. The participants verbally walk through the project from beginning to end, identifying the activities, their sequence, and their interrelationships. The scheduler then prepares the graphical schedule manually or by computer. The group meets again to review the first rough activity diagram. When agreement is reached on activities and logic, the scheduler obtains input that will affect the schedule from the owner, architect/engineer, subcontractors, equipment suppliers, and suppliers of long lead-time items. This information is incorporated into the rough diagram and reviewed again. If there are conflicts between the contractor's team and the information provided by the other parties, some negotiation or conflict resolution is in order. It is best to resolve any differences in opinion before the project starts. When all conflicts are resolved, the plan is accepted (signed-off). The rough diagram is ready for scheduling, which is the phase when durations, and possibly resources, are applied to the planned activities. During the scheduling stage, a certain amount of fine tuning is always necessary. Rough diagram planning must be reconciled with the estimate (cost and resource constraints) and time constraints (usually imposed by the contract documents).

List of Activities

In the second approach to the preparation of the rough diagram, the scheduler in the initial meeting makes a written list of activities instead of using a recording, remarking on the relationships between activities and other pertinent information.

Sketch for the Diagram

In the third approach, during the initial meeting the scheduler simply sketches the diagram (usually PERT) on a large, usually continuous, sheet of paper showing the activity names and logic. The disadvantage of this approach is that sometimes the scheduler gets so bogged down in diagramming that the flow of conversation and the effective use of meeting time are diminished. If the meeting has more than two people present, this method is usually not efficient.

Combination Approach. A combination approach that consists of listing the activities and simultaneously sketching a diagram is popular. After identifying the activities and discussing their relationships in the initial meeting, the planners write activity names on small stick-on notes. The notes can be moved around on a large sheet of paper by members of the scheduling team to refine the original logic in accordance with the accepted plan.

Reconciliation. In the first three approaches to creating a rough diagram, no reconciliation to the estimate is made until *after* the rough diagramming stage. Reconciliation is done during scheduling, when durations and resources are assigned to the project activities. All costs input to the estimate have to be input a second time if they are to be used in the schedule. This extra step of having to reenter all the information prohibits many contractors from using labor or resource profiles and detailed cost analysis by activity.

Software Integrator

The fourth approach—using a software integrator to produce the rough diagram—differs fundamentally from the others. This is the most efficient method for a number of reasons, primarily because it combines the estimate and the schedule simultaneously. The estimate is the document that defines the bottom line for a project—the budget for resources and costs. Unfortunately, the estimator and the scheduler look at the world differently. The estimator's goal is to turn out the greatest number of estimates with the best accuracy for the least cost. He or she will probably only get one out of every six to ten jobs estimated. The estimator is more concerned with defining cost than with providing a breakdown of cost for sequencing purposes. For example, an estimator will use one entry for concrete grade beams of a particular type with breakout by the different materials making up the grade beam (concrete, formwork, rebar, embeds, finishing, waterproofing, rubbing, etc.). This is the most efficient way to estimate cost. The scheduler, however, usually needs to relate the work by type of crew (carpentry, finishing, etc.) to individual pours or groups of pours. Since labor crews are commonly the resource to be maximized and kept working efficiently, the scheduler needs information sorted by type of crew.

Using an estimate/schedule software integrator for preparing the rough diagram avoids the problem of inputting the same information twice. Many of the best-selling estimating programs provide an interface program for the estimate to be "dumped" into the estimate/schedule integrator. While the information is in the integrator, individual line entries from the estimate can be either combined or split and attached to a named activity. The strength of the integrator is its ability to name the activities and move any associated cost and resources in the estimate to the named activity. The named activity retains the cost and resources when it is dumped to the scheduling software package without having to be reinputted, which ensures the correlation between the data in the estimate and in the schedule. The software integrator cannot be used to establish relationships, however. The relationships are established when the new file is brought up in *P3e*. This simplification process helps ensure that the contractor actually uses the management tools available through software packages such as *P3e* to control cost and resource functions.

Markup of the Estimate.   When using a software integrator approach to produce the rough diagram, the first step is for the project team to mark up the hard-copy printout of the estimate and define the activities manually. In the integrator, line entries from the estimate are either combined or split to form activities. When complete, the computer estimate file is dumped into the scheduling software package. All associated cost and resource information identified by activity is brought into the scheduling software package automatically when the dump occurs. The scheduling software package establishes relationships among the activities and calculates the schedule. This version of the rough schedule is presented to all key project participants for fine tuning, as are rough diagrams in the first three approaches. This approach to producing the rough diagram offers the advantage of reconciling the estimate and the schedule without having to input information twice, but care must taken not to let reliance on the computer stifle the creativity and freedom of the planning stage.

ROUGH DIAGRAM PREPARATION

There are five steps in the development of a rough diagram. The steps for preparation of a construction-only schedule are as follows:

1. **Contractor's Initial Meeting.**   The contractor's project team members have an initial meeting to discuss the entire project from beginning to end. The members include the estimator, project manager, possibly the superintendent, and the person creating the schedule. The team builds the project on paper, starting from the estimate. Either a tape recorder is used or the scheduler takes notes. Some people prefer to simply list the activities. In any case, the estimate is the benchmark from which to start.
2. **Rough Diagram Creation.**   The scheduler uses the information from the initial meeting to draw the rough diagram (usually PERT) on paper, showing interrelationships between the activities. Some schedulers prefer to go straight to the software at this stage rather than to manually draw the rough diagram on paper. Those who have software with an integrator function can also use it to estimate cost and resources needed for each activity at this stage.

3. **Review.** The entire project team must "buy into" the rough diagram so that it is "their" schedule. It is critical for each to agree as a team to the concept, methods, and procedures to be used in the construction of the project.

4. **Major Subcontractors and Suppliers Input.** Input from major subcontractors and suppliers of long lead-time material and/or equipment items is critical. With the contract documents, the general contractor or construction manager has overall responsibility for coordination and scheduling of the work. It is up to the contractor to coordinate all parties' work and to resolve any disputes. Input is usually required of the subcontractor in the contractor-subcontractor contract. Through discussions with the subcontractor/suppliers, the contractor modifies the plan to reconcile differences. Then the subcontractors/suppliers accept the contractor's plan as the project plan. Contractors need to make sure they have obtained information from any outside party that can impact the schedule.

5. **Project Schedule Acceptance.** Once the contractor has reviewed the revised rough schedule with the subcontractors/suppliers, it becomes the official project schedule.

Information Gathering for the Rough Diagram

During the prebid or estimating phase of the project life cycle, the estimate becomes the focal point in the gathering of information relating to the project. Information must be gathered from many sources and incorporated into the estimate. The following is typical of the information that must be collected:

- Owner time constraints and other input
- Scope definition
- Building methods and procedures to be used
- Productivity rates, crew balances, and crew sizes
- Labor availability and wage rates
- Construction equipment to be used
- Construction equipment availability and use rates
- Material availability and prices
- Subcontractor availability and prices
- Fabricator availability and prices
- Project organizational structure
- Rough preliminary schedule
- Temporary facilities requirements
- Permit and test requirements
- Tax requirements
- Insurance requirements

In-Depth Planning. If a contractor's bid wins the contract for a project, the estimate is used to generate the schedule. Now that the project is a real live job rather than just a proposal, more in-depth planning can take place. A logic diagram rather than a "quick and dirty" bar chart is used to fine-tune the information already gathered in the estimating stage.

Parties.   The schedule becomes the focal point of project information that is received from the project manager, superintendent, foremen, estimator, subcontractors, fabricators, suppliers, vendors, owner, and architect/engineer.

Meetings Held for the Preparation of the Rough Diagram

There are five important considerations to remember when holding meetings to produce the rough diagram. They are using an agenda, making a list of action items, preparing the project schedule, team building, and brainstorming.

Agenda.   To keep a meeting from degenerating into a general waste of time requires an agenda. Distribute the agenda prior to the meeting. It should document the date, time, and location of the meeting, information to be covered, what each participant should bring to the meeting, and points each participant should know before the meeting begins. Each member should already be familiar with the contract documents and the major parameters. It is a waste of everyone else's time if some project members come to the meeting unprepared.

Action Items.   The immediate product of a scheduling meeting is a list of agreed-upon action items to be accomplished before the next meeting. The list should designate the person(s) responsible for each item and set a date for the next meeting. The way we schedule and carry out the scheduling says a lot about our management abilities in carrying out the project.

Project Schedule.   It is frequently commented that the act of preparing the rough diagram is the most valuable part of the entire scheduling process. The reason is communications: Everyone is aware of the "plan"; everyone knows everyone else's point of view regarding project goals. The project schedule that is accepted after all discussion has taken place is the one used by all parties to organize their work.

Team.   If all key parties feel that they are a part of the team with a chance to contribute their information and ideas, the project is more likely to succeed. Some of the suggestions for improving project efficiency or lopping time off the schedule may come from unexpected sources. Communication leads to evaluation and reevaluation of the plan to streamline construction.

Brainstorming Alternatives. When a decision has to be made to choose a particular system or technique for some portion of the work, brainstorming is a good way to look at the alternatives. For example, to decide on the most effective way to pour the concrete columns in a multi-story building, the group would start by simply tossing out the question: "What is the most effective way to pour the concrete columns on this project?" Each member of the group replies by tossing out the first solution that comes to mind, no matter how ridiculous the solution might immediately sound. Try to get many possible solutions before any analysis of the solutions begins. After all proposed solutions have been introduced, discuss each one and eliminate those that obviously won't work. You should be able to narrow your list down to only two or three legitimate choices from which the group can make a final selection. The object of this brainstorming exercise is to break established patterns or mind-sets and look for new and better ways to build the project. Many times you get stuck in a rut doing things in the same old way simply because "that's the way we have always done it and it has always worked." Only by looking for new and more effective ways to build the "mouse trap" can we improve construction operation.

THE CONSTRUCTION PROJECT DEFINED

Most construction companies look upon each project as a profit center. The project represents a unique set of activities or actions that must take place to produce a unique product. The lump-sum project is divided into actions that precede and follow the signing of the contract. Precontract relates to the bidding. The contractor wants to invest as little as possible in the project until the client has signed the contract. Once the contract is formalized detailed planning and scheduling proceed. Only after the contract is in hand should the constructor commit to the level of detail and resources needed to produce a "production schedule." At the end of the project, the owner takes possession of the structure to use it for its intended purpose, takes responsibility for utilities and insurances, and submits final payment to the contractor.

Criteria for Success

Each project has a definable start and end and a unique set of characteristics and activities that must be accomplished to fulfill a contract. Each project is judged as a success or failure in terms of a preset list of criteria. For example:

- Did the project come in under budget?
- Did the project make any money?
- Were changes controlled through change orders?
- Did the project come in on time?

- Did the project meet the requirements of the contract documents?
- Were both client and employee satisfaction achieved?
- What was the project safety record?
- Was this project an effective use of company resources?
- Did the project meet, or exceed, the company's short- and long-term goals?

Uniqueness of Each Project

No two construction projects are quite the same. Even if the plans and specifications are almost identical, the projects' sites differ. The starting dates also differ, so each project is subject to different weather conditions. Environmental requirements, permits, and regulatory personnel differ from job to job. The project team that builds the project is usually also composed of different people. Each project is thus a learning experience. Lessons learned about productivity, project layout, flow of materials, crew sizing, and communications can help improve the quality of estimates and schedules in future projects.

PROJECT PHASES

Time and the control of time relates to all phases of the construction project life cycle. The construction project life cycle can be broken into eight phases:

1. Feasibility study
2. Conceptual design
3. Detailed design
4. Bidding
5. Construction
6. Commissioning
7. Closeout
8. Maintenance

1. **Feasibility Study.**   The feasibility study usually involves evaluating different design scenarios from the standpoint of building cost, maintenance cost, and appearance. Contractors use cost studies to play "What if?" games in an effort to pick the best approach. Only sketches are drawn, and only general concepts are considered at this stage. The great expense of producing detailed documents is not incurred until all major concepts have been agreed upon. Much of the project's costs are fixed when the feasibility study is complete. The reason these costs are fixed is that by this stage the following cost-determining characteristics of the project have already been defined:

   - Size of the project (number of square feet, number of floors)
   - Type of building system (steel building versus concrete building, flat slab versus waffle slab)

- General arrangement showing bathrooms, etc.
- Geographical location

The feasibility study phase is also where the global, high-dollar decisions about the project are generally defined and refined. The costs for implementing ideas are estimated and compared to the owner's budget and calculations of return on invested capital. Thus, during the feasibility study phase decisions about the economic viability of the project are made, before the owner invests in detailed plans and specifications and concrete. When participants believe that a viable project is attainable, conceptual design can begin.

2. **Conceptual Design.**   In the conceptual design phase, the design professional (architect/engineer) defines the owner's need in a conceptual project-scope document. This document defines the project in enough detail so that detailed design can begin. The conceptual design information needed varies by type of project. An office building differs from a refinery, for example, but the general idea is the same. The conceptual design of a new building would contain sketches of the following:

- Site
- Footprint of the building on site
- General floor plan of the building by floor
- Major wall sections
- Major elevations
- General definition of traffic flows
- Major environmental considerations, such as type of HVAC system
- Conceptual estimate (must include a projection of project duration)

3. **Detailed Design.**   Detailed design involves applying the broad concepts as defined in the conceptual design and producing the detailed plans, specifications, and the rest of the contract documents. To make a lump-sum bid specifying cost value and duration in days, a contractor has to know the scope of the project. The purpose of the detailed design is to produce the contract documents in enough detail so that they can be used for three primary purposes: bidding the project, building the project, and settling of any claims. A clear, concise set of contract documents is a tremendous asset to the constructor during the bidding, construction, and closeout phases of the project.

4. **Bidding.**   In the bidding phase of the project life cycle, the constructor prepares the estimate based on the contract documents prepared in the detailed design phase. The constructor quantifies the project in terms of material quantities, productivity rates, worker hours, wage rates, material dollars, subcontractor dollars, overheads, and indirect expenses. A rough schedule, usually a bar chart, is prepared for use in the bidding process.

5. **Construction.**   During building, the constructor marshals at the job site all the management, expertise, manpower, materials, equipment, subcontractors, temporary facilities, and financial wherewithal

necessary to construct the project according to the contract documents. The goal is to complete the project within cost and time constraints in order to make money.

6. **Commissioning.** Commissioning is the process of testing and start-up of systems. It includes the final inspection and preparation of a punchlist of remaining items to be completed, modified, or repaired before final acceptance by the owner. It is very important at this stage that the project be as nearly complete as possible. An attempt to close out the project when obvious items are incomplete may produce frustration and mistrust on the part of the owner and designer.

7. **Closeout.** The closeout phase in the project life cycle involves completing the paperwork necessary for a contractor to receive final payment and be released from the project. Documents include:

   - Affidavit of release of liens from suppliers, vendors, and subcontractors
   - Maintenance/owner's manuals for equipment and systems
   - As-built drawings
   - Request for final payment/lien waiver
   - Consent of surety
   - Guarantees
   - Warranties

8. **Maintenance.** Construction projects are designed for 20, 30, 50, or more years of productive life. To last that long, they require continual maintenance. Even on a project with a 50-year anticipated life cycle, the roof may only have a 20-year anticipated life. The elevator and the HVAC system require constant attention. Some contract proposal forms require from the contractor not only a bid for construction of the building but also for maintenance for a certain number of years of the building's life. Maintenance, however, is usually an ownership function.

PROJECT TEAM

No construction project of any size is ever built by an individual. It is constructed by a team. The concept of "team" crosses company boundaries. The makeup of the team preparing the rough diagram depends on the construction life cycle phases to be controlled with the schedule. Members of the lump-sum, construction-only scheduling team include the following:

Contractor's Organization
Project team
- Project manager
- Superintendent
- Foremen/craftspersons

Home office support
- Estimator
- Scheduler
- Cost accountant
- Purchaser
- Accountant
- Equipment manager
- Lawyer

Subcontractor's Organization

Field team
- Project manager
- Foremen/craftspersons

Home office support
- Estimator
- Scheduler
- Cost accountant
- Purchaser
- Accountant
- Equipment manager

Vendors'/Suppliers' Organizations

Fabrication/storage yard team
- Project manager
- Foremen/craftspersons

Home office support
- Estimator
- Purchaser
- Accountant
- Equipment manager

Owner

Construction representative

Architect/Engineer

Project representative

- Engineering consultants

Government

Inspector
Labor law enforcement official
Safety law enforcement official
Tax enforcement official

Successful projects require a mindset of teamwork and problem solving. The goal is to work together to settle disputes at the lowest level for the mutual benefit of the entire group, rather than engaging in fighting, finger pointing, and litigation. Once again the modern concepts of "partnering" and "team" cross company boundaries. Looking at the project from the perspective of a team rather than one dominated by company boundaries helps control time and project duration. The contractor is not solely in control of all the variables required for successful project completion. Only through teamwork can the project schedule be successfully met. One of the best ways to ensure that teamwork is effective during the entire construction process is through thorough and accurate written project documentation. Many decisions made during the project and sealed with a handshake are forgotten later on unless written documentation formalizes the decisions reached.

DESIGN-BID-BUILD CONSTRUCTION

Design-bid-build construction carries out the construction life cycle phases in consecutive rather than concurrent, order—each phase is completed before the next begins. Building the project with this constraint takes longer. Since each step of the process is defined, finished, and usually paid for before the next begins, and since this process is usually based on lump-sum contracts, the owner transfers the risk of cost overruns to other parties. The owner knows the cost of each phase before committing to paying for it. The owner also knows that because the project can be canceled at any time, he or she is only at risk for the work that has been released.

No Contractor Input

In a lump-sum contract (design-bid-build construction), the contractor's contract typically includes the construction, commissioning, and close-out phases of the construction life cycle but no input during the conceptual or detailed design phases. This is unfortunate, since who knows more about minimizing construction cost by efficient design, constructibility, materials selection, and the use of prefabricated materials than someone who is involved in the actual building process every day? Many times the designers operate in a vacuum during the design phase relying only on their own past experience to select materials and methods of construction. Significant savings can usually be realized by involving contractors, subcontractors, and material suppliers at this stage of the project. The use of construction management professional services may be a good solution, as it incorporates the construction experience in the design process, before a contractor is available.

Steps in the Lump-Sum Project

The following is a sequential listing of the steps typically followed with lump-sum construction in the United States:

1. Owner determines a need.
2. Owner contacts architect/engineer for conceptual design.
3. Conceptual design is completed.
4. Owner approves conceptual design or sends it back for modification.
5. Owner contracts architect/engineer for detailed design contract documents.
6. Detailed design is completed.
7. Owner approves detailed design or sends it back for modification.
8. Owner puts contract documents out for bid.
9. Bids are received from contractors and negotiated.
10. Owner and contractor sign contract.
11. Construction proceeds.
12. Project is commissioned.
13. Project is closed out.

Negotiated, Fast-Track Construction

With a fast-track type project, the constructor (construction manager) is typically involved in all construction life-cycle phases except possibly maintenance. The constructor has a negotiated contract with the owner and acts as the owner's agent in a fiduciary relationship. He or she is looking out for the owner's best interest. The constructor has input in defining the owner's needs during both the conceptual and detailed design phases. The contractor is primarily concerned with minimizing construction cost by efficient design, constructibility, materials selection, and the use of prefabricated materials while the critical decisions about these factors are being made. The constructor will produce the conceptual estimate and cost studies of different design scenarios. Having a member of the design team who is a cost-conscious, knowledgeable constructor can be a tremendous advantage to the owner in producing a successful project.

Steps in Fast-Track Construction

The following set of steps is typical of fast-track construction in the United States using the agency construction management type contract:

1. Owner has a need.
2. Owner contacts construction manager for services, which include control of design, estimating, and scheduling. The construction manager may be the architect or engineer as well as the contractor.

3. Construction manager works with architect/engineer to produce conceptual design.
4. Construction manager estimates costs to fine tune conceptual design until it meets the owner's return-on-investment requirements.
5. Owner/construction manager approves conceptual design.
6. Owner/construction manager contracts with architect/engineer for detailed design contract documents, then breaks down design into packages or phases that can be completed and put out for construction bid.
7. Owner/construction manager approves detailed design one package at a time.
8. Owner/construction manager puts contract documents out for bid, one package at a time.
9. Owner and successful package contractor sign contract for single package.
10. Owner/construction manager brings all the packages through the steps of detailed design, bid, award, and construction.
11. Construction progresses simultaneously with further detailed design. The construction manager monitors the schedule, controls multiple contractors, and approves pay requests. The construction manager acts as the general contractor at the job site.
12. Project commissioning is usually handled one package or system at a time.
13. Project closeout is handled by the construction manager in much the same way as the general contractor would handle it.

With the fast-track approach to construction services, the owner assumes more of the risk, since construction is proceeding before the design documents are completed. The construction manager is the owner's agent, and therefore typically does not sign a lump-sum contract with the owner., unless it is a CM (Construction Management) at risk contract. The real advantage to the owner is the time savings of producing the detailed design and construction concurrently rather than consecutively. This approach can cut in half the overall time needed to bring a project "on line," thus saving not only time but money.

USE OF SCHEDULES

A critical part of any scheduling effort is in deciding how the schedule will be used to control a project. The following issues should be defined before the rough diagram is prepared:

- Type of project
- Purpose of the schedule
- Software requirements

- Parties involved: owner, construction manager, architect/engineer, contractors, major subcontractors, minor subcontractors and sub-subcontractors, suppliers of long lead-time items
- Authority/responsibility of involved parties (Who is "keeper of the schedule"?); responsibility for resolving disputes and interferences
- Needs of the involved parties from a scheduling point of view
- Type, frequency, and depth of reports to be provided
- Schedule update requirements (What is the time frame for providing updated information in what format, and who is to provide the information?)
- Project change requirements
- Resources to be controlled: labor, materials, subcontractors, construction equipment
- Cash flow requirements
- Payment (schedule of values) requirements

TIME UNITS

The Day

The time unit used for most construction schedules is the day. *P3e* does not have a concept of time units. Units are strictly for display purposes only.

The determination of the minimum planning unit depends on the total expended duration of the project, as well as the required level of detail. In the typical construction schedule with days as the unit of measure, when activities are being defined, no block of work is assigned a duration of less than a day to complete. If the activity takes less than a day, it is either rounded up to a day or combined with another task/activity.

The Hour

Sometimes the hour is a more convenient unit than the day. This is the usual unit of choice for a "turnaround" schedule as used in a paper mill, chemical plant, refinery, or other facility in operation. The facility must be brought down or "off-line" for maintenance or to add or modify plant systems. Typically, the owner loses thousands of dollars per day while the plant is out of operation. Making the modifications as quickly as possible and getting the plant going again is essential to the owner. Each hour is critical.

CONCURRENT RATHER THAN CONSECUTIVE LOGIC

Besides trying to improve efficiencies, schedulers also need to improve the logic of activity interrelationships and sequences when putting the rough diagram together. Instead of having just one thing at a time happen at the job site, as many things, crews, or work functions as possible need to be happening, without their interfering with each other and without risking exposure or damage by project components being in place too early. What is the shortest and most efficient way to accomplish the work?

ESTIMATING ACTIVITY DURATION

The duration of an activity is a function of the quantity of work to be done by the activity and the rate of production at which the work can be accomplished. The formula is:

$$\text{Activity Duration} = \frac{\text{Quantity of Work}}{\text{Productivity Rate}}$$

Take, for example, a masonry activity with a quantity of work of 10,000 regular masonry blocks to be placed by a crew of three masons, two laborers, and a mixer. The crew can place 800 blocks per day. Thus, duration = 10,000 blocks/800 blocks per day = 12.5 or 13 days.

RESOURCE AVAILABILITY

Critical to estimating activity duration is the availability of resources needed, including labor, materials, equipment, subcontractors, and suppliers/fabricators.

Driving Resources

Most activities have certain driving resources that control activity duration. For example, in the preceding masonry activity, the three masons are the driving resource. Their craft determines activity duration. Once the ratio of laborers and mixers has reached maximum efficiency for three masons, no matter how many more laborers and mixers are added, the performance of the three masons will not be increased.

Reduction of Duration

There are three ways to reduce the duration of 13 days for the masonry activity. The first is, by increasing the productivity rate of the three masons to better than 800 blocks per day. This could be accomplished by an improved method or system for placing the block. The second way would be to add more masons, with enough laborers and mixers to ensure full production. The third way would be to have the masons work overtime (more hours per day).

Extended Scheduled Overtime

Extended scheduled overtime can ultimately have a negative impact because of lost productivity. It is also important to consider the higher cost of overtime pay. Studies have shown that consistent reliance on overtime by the same crew actually results in less productivity than that normally achieved in an eight-hour day.

As a rule, the more resources that are available to put the activity in place, the less time it takes to place the activity. If any of the resources necessary to place the activity are missing or are not handled properly, the activity duration calculation is impacted.

QUANTITY OF WORK

The contract documents define the scope of the project. Using the plans and specifications, the estimators survey quantities of materials necessary to complete the project. The estimate may not provide a bill of materials in enough detail to enable purchasing of all materials for the project, but it will provide enough detail to bid the project. The estimate prepared at this stage relies solely on the contract documents. If insufficient detail or ambiguity exists in the contract documents, the estimate will usually be overly conservative to account for uncertainties.

To produce a *quantity survey*, the estimators organize by type of work. Building contractors usually use Construction Specifications Institute (CSI) cost code structure. The estimators assign work definition by spec division, by phase, then by the item or unit of work within the phase.

Contractors usually produce a quantity survey only for work to be completed by the contractor's own forces. For subcontracted work, contractors usually depend on the market for the best price.

A code of accounts is a coding/numbering system to categorize work for controlling/tracking purposes. The contractor organizes the code of accounts and quantity survey that each type of crew will accomplish into identifiable areas or phases, usually organized according to CSI format. The major CSI phase headings are:

01000	General Requirements	09000	Finishes
02000	Sitework	10000	Specialties
03000	Concrete	11000	Equipment
04000	Masonry	12000	Furnishings
05000	Metals	13000	Special Construction
06000	Wood and Plastics	14000	Conveying Systems
07000	Thermal and Moisture Protection	15000	Mechanical
08000	Doors and Windows	16000	Electrical

The cost code structure is organized by craft designation so that work boundaries are understood by all parties involved. For example, all masonry-related work is categorized in the 04000s.

PRODUCTIVITY RATE

Productivity Rate by Activity

The quantity survey defines the amount of materials for a particular type of work, such as the square footage of a slab to be finished. The next step is to assign a productivity rate to determine the duration and resources necessary for the activity. The rate is the quantity of work accomplished per work hour (e.g., 30 square feet of concrete finished per work hour, or 30 SF/WH), or quantity of work accomplished per crew hour (e.g., 120 square feet/crew hour based on a four-person crew, or 120 SF/CH). The rate per man-hour (work hour) has to be adjusted for the size of the crew to determine duration.

Sources of Productivity Information

Sometimes if no formal estimate is prepared, the scheduler may have to come up with the productivity rate. The following sources may be useful:

• Company records for producing the same type of work on previous projects (historical data)
• Published information in reference text
• Observation and measurement of work performance as it is being put in place on another project
• Qualified expert opinion

FACTORS AFFECTING PRODUCTIVITY

The amount of time it takes to accomplish a unit of work can vary appreciably for the same type of work from project to project and is the reason for unpredictability of construction labor costs. This variability in performance results from differences in communications, supervision/ proper preplanning, layout of the work, crew balance, skill/ craftsmanship, mental attitude of workers, purchasing practices, working conditions, environmental factors, continuously scheduled overtime, safety practices, work rules, and availability of work.

Communications

Good communications—in writing—cannot be stressed enough when discussing productivity. It is common practice for the superintendent to communicate orally with the foreman about the work to be accomplished, referring to the plans, specs, and shop drawings. The foreman in turn communicates orally to craftspersons the direction, methods, and layout for accomplishing the work. This chain of oral communication leads to a great deal of rework, exemplified by the comment "I built what I thought you wanted."

Some supervisory personnel seem to almost take pride in not communicating information to their subordinates. This attitude of "I'm the only one who knows all the answers" promotes distrust and low morale among employees. The effective supervisor, however, has learned that open and effective communications with everyone involved in the construction process makes his or her job much easier and cuts down on needless finger pointing. A good subconscious thought for this manager to constantly ask is "If the owner of this project came onto the job site and asked any one individual what they were working on (and why), would that person know the answer?"

Many contractors now put communications in writing, using 10-day look-ahead and job-assignment sheets for in-depth planning and scheduling. An emphasis on sketches and drawings further reduces dependency on oral communications. The improved documentation improves productivity.

Supervision and Preplanning

Good supervision and proper preplanning also improve productivity. Commitment to cost and schedule control by top management leads to improvements in training, safety, scheduling, estimating, and purchasing and to an emphasis on quality on all projects.

Efficient Layout of the Work

Of critical importance in the proper execution of the work is efficient layout. These include:

- Storage of materials so they can be located quickly and easily when necessary

- Minimal handling of materials (on many work items, more time is spent handling the materials than actually putting the materials in place)
- Use of the right equipment for the job
- Efficient access to tools, utilities, drinking water, and sanitary facilities
- Where possible, working at waist level to reduce unnecessary motion and fatigue

Proper Crew Balance

A construction crew has a proper balance of workers for various aspects of each activity. In a masonry crew, for example, the ratio of the workers mixing the mortar to the laborers transporting the mortar and stacking the block for placement and to masons actually placing the block must be balanced properly to efficiently accomplish the work. The ratio of highly paid skilled workers to the lower paid helpers is important for peak efficiency.

Skill/Craftsmanship

Training construction craftspersons to improve their skills is necessary because of the ever-changing technology. Improvements in construction equipment are changing the way work is accomplished and increasing productivity. Since a construction company's primary asset is its employees, nurturing and training them to increase performance is a wise investment.

Mental Attitude of Workers

People are not machines. The mental attitude of workers seriously affects productivity. Management should strive to promote worker feelings of pride in the company, faith in a secure future, confidence in good management, the company's respect for employees, the company's regard for employee's opinions, company growth, and the potential for individual growth. When employees get satisfaction from working with a company and their individual needs are being met, they are more productive at the job site.

One way that many construction companies have discovered to improve the attitude of their workers is by adoption of a "Total Quality Management" (TQM) philosophy. This organizational and operational approach, originally implemented in the manufacturing sector, has proven to be beneficial for the construction industry as well. The Construction Industry Institute in a study of 17 major contractors who had developed a TQM operating philosophy found they had four major things in common:

- Greater productivity
- Increased customer satisfaction

- More profits
- Higher employee morale

As well as the obvious benefits of increased morale, many construction companies have documented savings in the form of better safety records and a significant reduction in "reworks."

Purchasing Practices

A construction company's purchasing practices can have a tremendous impact on job site productivity. The use of prefabricated assemblies or preassembled units or taking labor off-site to a manufacturing-type controlled environment can have great impact on worker hours spent at the job site. A lower average wage rate, lower-level craft skills required, increased availability of specialized tools and jigs, the ability to work under a roof in a controlled environment, and availability of local labor can all be great advantages. A scheduling advantage is that the fabrication can be concurrent with work at the job site, reducing the overall time for the project to be completed. Prefabricating materials in a controlled environment rather than at the job site can also have a huge impact on cost.

Working Conditions

In construction, the worker is typically exposed to the elements. Extremes can slow productivity. Conditions that influence job site productivity include some that cannot be controlled—heat, cold, rain, humidity, dust, wind, odor, and acts of God—and some that can be controlled—noise, climbing up or down, bending low or reaching high, and the number of people on site.

Some companies have taken innovative approaches to address the day-to-day working conditions of their employees. Texas Instruments Corporation (TI), for example, initiated radical changes to the way construction was carried out during the building of many of their newer facilities. Texas Instruments implemented a TQM policy to be followed by all of the general contractors and subcontractors working on TI projects. One of the requirements was the construction of a clean, air-conditioned lunchroom for the workers at every one of their job sites. The only stipulation was that no food or drink (other than water) was allowed on the actual construction site. This one change, along with other elements of the TQM philosophy, resulted in higher employee morale, increased profits reported by all of the contractors involved, and a reduction in "reworks" from 10% to 2%.

Continuously Scheduled Overtime

Many studies have shown that continuously scheduled overtime has a disastrous impact on productivity. The construction industry Business

Round Table study entitled "Cost Effectiveness Study C-3" (November, 1980) showed that workers putting in 60-hour weeks for nine straight weeks accomplished the same amount of work in the ninth week as in a normal 40-hour workweek without the overtime. This means that paying for the extra 20 hours accomplished nothing. Compounding this false economy is the enormous cost of paying overtime for work in excess of 40 hours per week. Spot overtime can be effective, but continuously scheduled overtime is decidedly ineffective from an economic point of view.

Safety Practices

Safety practices have direct and indirect influences on productivity. The direct influence is that when someone is hurt, work at the job site usually stops or is at least impacted by the disturbance. It is the topic of conversation. Everyone is concerned. The company has at least temporarily lost the services of an employee. The loss also affects the crew balance and hence the productivity of a work unit.

The indirect impact relates to the effect accidents have on workers' morale and feelings of personal safety and security. Accidents affect the way employees think of the quality of the company's management and ability to manage the job site. Accidents also affect the employee's pride in the company and desire to stay with the company.

Work Rules

Sometimes, labor constraints can have a negative impact on productivity. Union constraints that may limit management's ability to organize for maximum productivity include jurisdictional disputes, production guidelines, limits on time studies, limits on piecework, and limits on prefabrication.

Availability of Work

If times are good and there is plenty of construction work available, employees know that other jobs are readily available. There is not as much pressure to produce at the existing job. Conversely, if things are tight, there is much more pressure to keep the existing job, or possibly to draw out the current work as long as possible. Work availability can be a double-edged sword, depending on how it is perceived by the workers.

ACCURACY OF ESTIMATING ACTIVITY DURATION

Schedulers sometimes make subjective judgments about the duration of activities rather than taking the time to refer back to the estimate and perform an actual calculation based on productivity rates and quantity of work. Since time is money, the accuracy of a contractor's schedule depends directly on the reliability of estimates of activity durations.

EXAMPLE PROBLEM: ROUGH MANUAL LOGIC DIAGRAM

Table 2–2 is a list of 28 activities (activity ID and description) for a house put together as an example for student use (see the wood-framed house drawings in the Appendix). The rough manual logic diagram (Figures 2–3a and 2–3b) was constructed using the list provided in Table 2–2.

Act ID	Act Description
A1000	Clear Site
A1010	Building Layout
A1020	Form/Pour Footings
A1030	Pier Masonry
A1040	Wood Floor System
A1050	Rough Framing Walls
A1060	Rough Framing Roof
A1070	Doors and Windows
A1080	Ext Wall Board
A1090	Ext Wall Insulation
A1100	Rough Plumbing
A1110	Rough HVAC
A1120	Rough Elect
A1130	Shingles
A1140	Ext Siding
A1150	Ext Finish Carpentry
A1160	Hang Drywall
A1170	Finish Drywall
A1180	Cabinets
A1190	Ext Paint
A1200	Int Finish Carpentry
A1210	Int Paint
A1220	Finish Plumbing
A1230	Finish HVAC
A1240	Finish Elect
A1250	Flooring
A1260	Grading & Landscaping
A1270	Punch List

Table 2-2   Activity List—Wood-Framed House

EXERCISES

Exercises 1 to 6 contain logic statements. Complete a logic diagram for these statements.

1.

1.	A	Must Precede	B, C, D
2.	B	Must Precede	E
3.	C, D	Must Precede	F
4.	E	Must Precede	G, H
5.	F	Must Precede	H
6.	G, H	Must Precede	I

Figure 2–3   Exercise 1—Logic Statements

2.

1.	A	Must Precede	B, C
2.	B	Must Precede	D, E
3.	C	Must Precede	F
4.	D	Must Precede	G
5.	E, F	Must Precede	H
6.	G	Must Precede	I, J
7.	H	Must Precede	J
8.	I, J	Must Precede	K

Figure 2–4   Exercise 2—Logic Statements

3.

1.	A	Must Precede	B, C, D
2.	B, C	Must Precede	E
3.	C, D	Must Precede	F
4.	D	Must Precede	G
5.	E, F	Must Precede	H
6.	G, F	Must Precede	I
7.	H, I	Must Precede	J
8.	I	Must Precede	K
9.	J, K	Must Precede	L

Figure 2–5   Exercise 3—Logic Statements

4.

1.	A	Must Precede	C, D
2.	B	Must Precede	D, E
3.	C	Must Precede	F
4.	D	Must Precede	G
5.	E	Must Precede	H, I
6.	F, G	Must Precede	J
7.	G, H	Must Precede	K
8.	I	Must Precede	L
9.	J, K	Must Precede	M
10.	K	Must Precede	N
11.	K, L	Must Precede	O
12.	M, N	Must Precede	P
13.	O	Must Precede	Q
14.	P, Q	Must Precede	R

Figure 2–6   Exercise 4—Logic Statements

5.

1.	A	Must Precede	B, C
2.	B	Must Precede	D, E
3.	C	Must Precede	F
4.	D	Must Precede	G, H
5.	E, F	Must Precede	I
6.	F	Must Precede	J, K
7.	G, H, I	Must Precede	L
8.	I	Must Precede	M
9.	J, K	Must Precede	N
10.	L, M	Must Precede	O
11.	M, N	Must Precede	P
12.	O, P	Must Precede	Q
13.	P	Must Precede	R

Figure 2–7   Exercise 5—Logic Statements

6.

1.	A	Must Precede	C, D
2.	B	Must Precede	E
3.	C	Must Precede	F, G
4.	D	Must Precede	G, H
5.	E	Must Precede	H, I, J
6.	F	Must Precede	K
7.	G	Must Precede	K, L, M
8.	H	Must Precede	L, M, N
9.	I	Must Precede	M, N
10.	J	Must Precede	O
11.	K	Must Precede	P, Q
12.	L	Must Precede	R
13.	M	Must Precede	Q, R
14.	N	Must Precede	R, S, T
15.	O	Must Precede	S, T
16.	P	Must Precede	U
17.	Q	Must Precede	V
18.	R	Must Precede	V, W
19.	S, T	Must Precede	W, X
20.	U	Must Precede	Y
21.	V	Must Precede	Y, Z
22.	W	Must Precede	Z
23.	X	Must Precede	A'
24.	Y, Z	Must Precede	B'
25.	A'	Must Precede	C'

Figure 2–8   Exercise 6—Logic Statements

7. **Small Commercial Concrete Block Building**
   Prepare a rough manual logic diagram for the small commercial concrete block building located in the Appendix. Follow these steps:
   A. Prepare a list of activity descriptions (minimum of 60 activities).
   B. Establish activity relationships.
   C. Create the rough manual logic diagram.

8. **Large Commercial Building**
   Prepare a rough manual logic diagram for the following large commercial building located in the Appendix. Follow these steps:
   A. Prepare a list of activity descriptions (minimum of 150 activities).
   B. Establish activity relationships.
   C. Create the rough manual logic diagram.

Section 2

Scheduling

Schedule Calculations

Objectives

Upon completion of this chapter, you should be able to:

- Calculate forward pass
- Calculate backward pass
- Calculate total float
- Calculate free float
- Complete a data table
- Correlate ordinal and calendar days

DEFINITIONS OF IMPORTANT TERMS

The mathematical calculations of a precedence (activity-on-node) diagram follow a few simple rules. To make these calculations, several terms and their definitions need to be understood.

Activity. A task activity has five specific characteristics. It consumes time, consumes resources, has a definable start and finish, is assignable, and is measurable. Construction projects are made up of a number of individual activities (events) that must be accomplished to complete the project.

Node. The node is the box (or other shape) containing activity information. Nodes contain varying amounts of information per the requirements of the diagram. The nodes are connected by arrows showing relationships among the activities. The nodes used in the diagrams in this chapter contain activity description, duration, early start, late finish, and float (Figure 3–1).

Precedence (Activity-on-Node) Diagrams. The nodes are the activities, and they are connected by arrows that show relationships among activities. The nodes contain information about the activities. This information may contain any or all of the following: identification, description, original duration, remaining duration, early start, late start, early

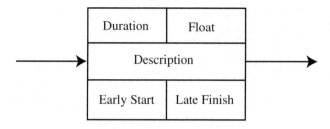

Node configuration used in this book

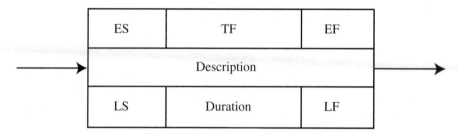

One of the most common node configurations used by *P3e*

Figure 3–1 Node Configurations

finish, late finish, and total float. Since *P3e* is written to support activity-on-node diagramming, this is the diagramming method used in this book.

Critical Path. The critical path is the continuous chain of activities with the longest duration; it determines the project duration. It should be noted that the critical path can change throughout the duration of a project. Activities that were originally secondary at the start of the project can become critical if they are delayed or exceed their total float. A non-critical activity can exceed its planned duration if it is within its total float without becoming critical.

Forward Pass. Calculations for the forward pass start at the beginning of the diagram and proceed forward to the end. They determine early start, early finish, and project duration. For a finish-start relationship, all predecessor activities must be complete before an activity can start.

Early Start. This is the earliest possible time that an activity can start according to relationships assigned.

Early Finish. This is the earliest time that an activity can finish and not prolong the project.

Backward Pass. Backward pass calculations start at the end of the diagram and work backward to the beginning, following logic constraints. They determine the late finish and late start of each activity.

Late Finish. This is the latest time that an activity can finish and not prolong the project.

Late Start. This is the latest time that an activity can start and not prolong the project.

Float. The amount of time difference between the calculated duration of the activity chain and the critical path is called float. It permits an activity to start later than its early start and not prolong the project. Float is sometimes called slacktime and may be classified as total or free.

Total Float. Total float is the measure of leeway in starting and completing an activity. It is the number of time units (hours, days, weeks, years) that an activity (or chain of activities) can be delayed without affecting the project end date. It is a shared property among activities on a certain path or chain.

Free Float. Free float is also referred to as activity float because, unlike total float, free float is the property of an activity and not the network path of which an activity is part. Free float is the amount of time the start orfinish of an activity may be delayed without delaying the start of a successor activity (depending on the relationship).

FORWARD PASS

Calculating Early Start

To calculate an activity's early start (forward pass), add the activity's duration to the early start of the preceding activity. All predecessor activities to the activity must be complete. Activity R (Figure 3–2) is the first activity and therefore has no predecessors. It therefore has an early start of 1 (Figure 3–3). Starting with day 1 produces the correct schedule duration plus 1. This simply means the last day of the project is used up and the calculation proceeds to the beginning of the day after the project is finished. The early start of G is 5 (Figure 3–3), or 1 (early start of R, the preceding activity) plus 4 (duration of the R, the preceding activity).

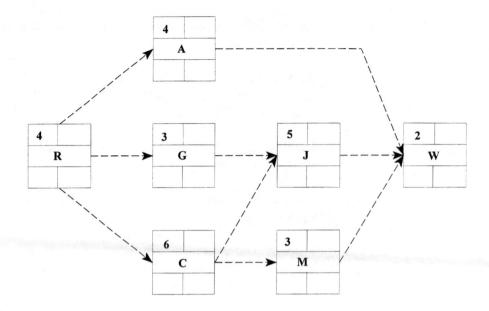

Figure 3–2   Precedence Diagram—Durations

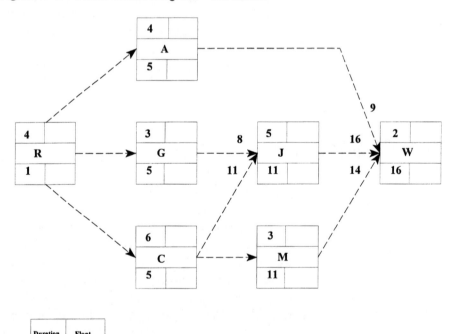

Figure 3–3   Forward Pass—Early Starts

Larger Value Takes Precedence in the Early Start

Where two chains converge on a single activity, such as Activity J (Figure 3–3), the larger early start value controls the path, since that path has the longer duration and *all* prior activities must be finished. The two choices are 8 days from Activity G and 11 days from Activity C. The 11 controls since it is the larger value. See Table 3–1 for the early start calculations for Figure 3–3. The early start formula is:

$$\text{Early Start} = (\text{Highest Value of Predecessor Early Start}) + (\text{Predecessor Duration})$$

BACKWARD PASS

Calculating Late Finish

The backward pass is used to determine activity late finishes. As the name implies, calculations begin at the diagram end and pass in a backward direction to the beginning. The formula for the late finish of an activity is the late finish of the following activity minus the duration of the following activity. The late finish of activity W in Figure 3–4 is the end of day 18. Since W is the last activity, it has to be on the critical path. W has an early start of 16 and a duration of 2. W has a late finish of 18 (16 + 2). The late finish of J is 16, or 18 (late finish of following activity) minus 2 (duration of following activity).

Activity	Preceding Activity	Early Start of Preceding Activity	Duration of Preceding Activity	Early Start of Activity (*Controls)
R				1
A	R	1	4	5
G	R	1	4	5
C	R	1	4	5
J	G	3	5	8
	C	6	5	11*
M	C	6	5	11
W	A	5	4	9
	J	11	5	16*
	M	11	3	14

Table 3–1   Early Start Calculations

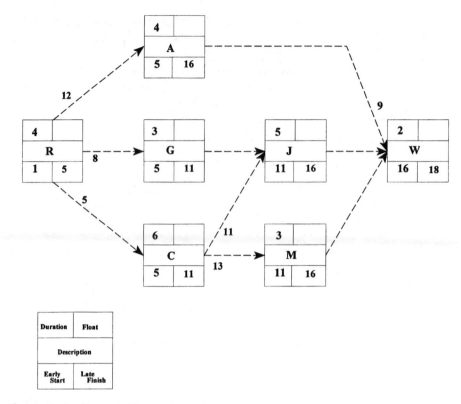

Figure 3–4   Reverse Pass—Late Finishes

Smaller Value Takes Precedence in Late Finish

When two paths converge on a single activity, the smaller value is used. The backward pass to C from the path through J has a late finish of 11 (16 – 5). The path to C through M has a late finish of 13 (16 – 3). Since 11 is the smaller value, it is used as the late finish of C. See Table 3–2 for the late finish calculations for Figure 3–4. The early late finish formula is:

$$\text{Late Finish} = (\text{Lowest Value of Successor Late Finish}) - (\text{Successor Duration})$$

FLOAT

Construction schedulers use the float to determine which activities are critical and which activities have "slack" or "fluff."

Flattened requirements float associated with noncritical activities can be used to "flatten" or level the requirements for resources, including personnel, materials, construction equipment, and cash. For example, if two

Activity	Following Activity	Late Finish of Following Activity	Duration of Following Activity	Late Finish (*Controls)
W				18
M	W	18	2	16
J	W	18	2	16
C	M	16	3	13
	J	16	5	11*
G	J	16	5	11
A	W	18	2	16
R	C	11	6	5*
	G	11	3	8
	A	16	4	12

Table 3–2   Late Finish Calculations

activities have a scheduled early start of the same day, and both require a carpentry crew, completing the critical activity first reduces the total number of carpenters needed. Such sequencing can similarly reduce material, equipment, and cash flow requirements. Noncritical activities do not have the same priority as critical activities and do not determine the critical path (project duration). It is important to note that if the float is used up, the critical path changes and previously noncritical activities may become critical.

Formulas to Calculate Float

The formulas for the calculation of total float require late start and early finish calculations. The formulas are:

Late Start = Late Finish – Duration
(Determined from backward pass)

Early Finish = Early Start + Duration
(Determined from forward pass)

There are three formulas for the calculation of activity total float:

Total Float = Late Finish – (Early Start + Duration)

Total Float = Late Start – Early Start

Total Float = Late Finish – Early Finish

Given the information contained in the nodes in Figure 3–5, the first formula is the most suitable. All three of these formulas must yield the same

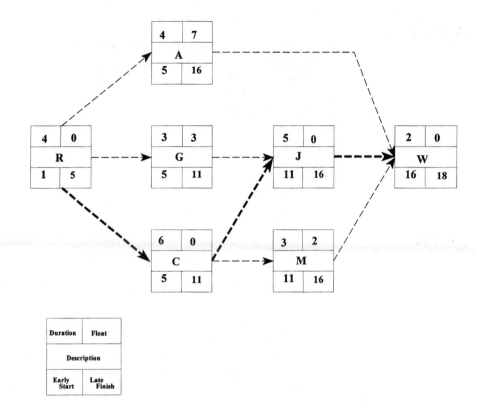

Figure 3–5 Float

value, or a mistake has been made. See Table 3–3 for total float calculations for Figure 3–5. By examining Table 3–4 for the example problem, you can see that the three formulas return the same value. The above three formulas for calculating total float hold true unless the activity is in physical progress. The float calculation for activities that are partially completed will be covered in the chapters on activity updating.

Activity	Late Finish	Early Start	Duration	Total Float
R	5	1	4	0
A	16	5	4	7
G	11	5	3	3
C	11	5	6	0
J	16	11	5	0
M	16	11	3	2
W	18	16	2	0

Table 3–3 Total Float Calculations

DATA TABLE

Table 3–4 is a classic scheduling data table (tabular sort) sorted by early starts. The table includes late starts and early finishes. Again the formulas for late starts and early finishes are:

Late Start = Late Finish – Duration
Early Finish = Early Start + Duration

CALENDARS

So far in this chapter, the scheduling units have been ordinal (numeric). To be more useful, ordinal days may be converted to calendar days (Figure 3–6). Calendar days account for nonworkdays, including Saturdays, Sundays, holidays, and a possible allowance for rain or other bad weather. These calendar decisions must be made to convert from ordinal days to calendar days (Figure 3–7).

After allowing for nonworkdays, the ordinal days can now be correlated with calendar days. The project starts on the tenth day of the first month (Figure 3–8).

P3e allows the user to assign an activity to any one of a number of possible calendars. Having different calendar days available for progress changes the ordinal to calendar day correlation. The concept of changing (having multiple) calendars will be covered in Chapter 4.

Activity	Duration	Early Start	Late Start	Early Finish	Late Finish	Total Float
R	4	1	1	5	5	0
A	4	5	12	9	16	7
G	3	5	8	8	11	3
C	6	5	5	11	11	0
J	5	11	11	16	16	0
M	3	11	13	14	16	2
W	2	16	16	18	18	0

Table 3–4   Data Table

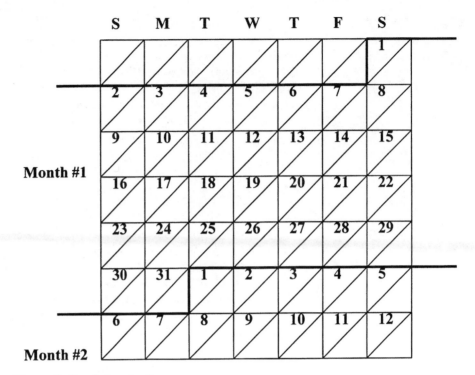

Figure 3–6   Calendar Days

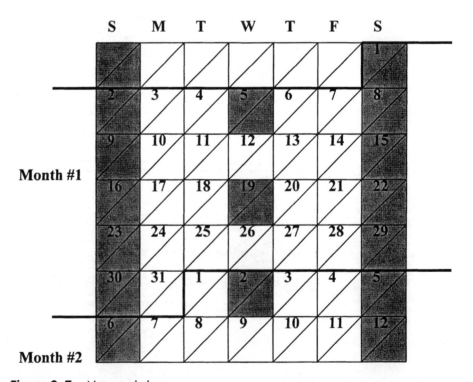

Figure 3–7   Nonworkdays

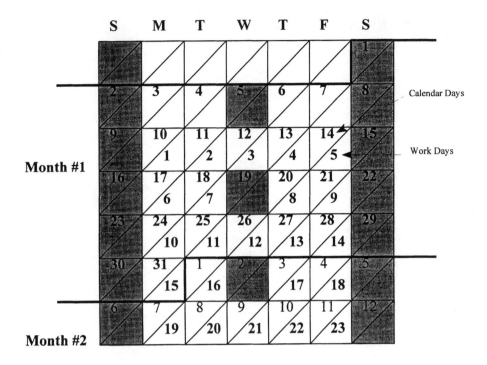

Figure 3–8 Work/Calendar Days

P3e CALCULATIONS

P3e calculates the forward pass and early starts as described in this chapter. There is a difference, however, in the way *P3e* performs the backward pass and late finish calculations from those described earlier. We have used the method described in this chapter because of its simplicity for the new user.

P3e calculates late finish in the following way. If you start an activity on day 1 and it takes 2 days to finish the activity, you would complete the activity at the end of day 2. Whereas early starts are based on starting at the beginning of the day, the late finish dates are based on the end of the day. To make this concept work, the *P3e* formulas are:

Early Start = (Highest Value of Predecessor Early Finish) + lag + 1
Late Finish = (Lowest Value of Successor Late Finish) – lag – 1
Late Start = Late Finish – Duration + 1
Early Finish = Early Start + Duration – 1

Lag in these formulas shows the amount of delay in a relationship.

EXAMPLE PROBLEM: CALCULATIONS

Figure 3–9 illustrates the forward pass, backward pass, and float calculation. Table 3–5 shows the same information in tabular format.

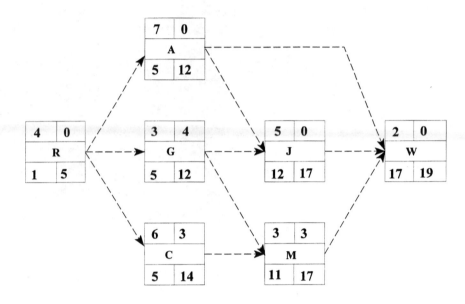

Figure 3–9   Example Problem—Precedence Diagram

Activity	Duration	Early Start	Late Start	Early Finish	Late Finish	Total Float
R	4	1	1	5	5	0
A	4	5	5	12	12	0
G	3	5	9	8	12	4
C	6	5	8	11	14	3
J	5	12	12	17	17	0
M	3	11	14	14	17	3
W	2	17	17	19	19	0

Table 3–5   Example Problem—Data Table

EXERCISES

Complete the precedence diagrams and data tables for Exercises 1 to 8.

1.

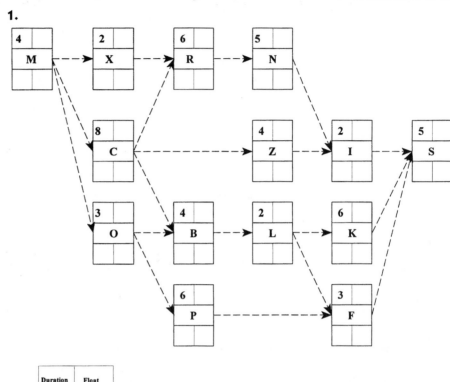

Figure 3–10   Exercise 1—Precedence Diagram

Activity	Duration	Early Start	Late Start	Early Finish	Late Finish	Total Float
M						
X						
C						
O						
R						
B						
P						
N						
Z						
L						
I						
K						
F						
S						

Table 3–6   Exercise 1—Data Table

2.

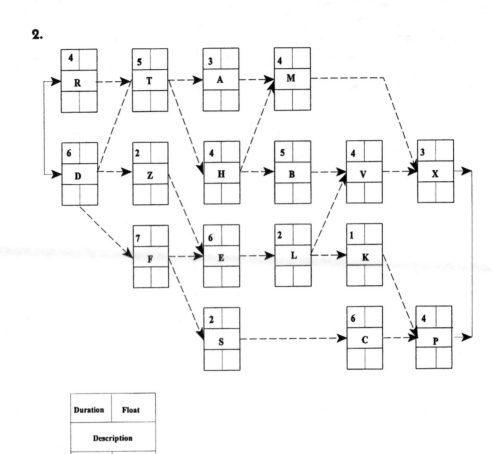

Figure 3–11 Exercise 2—Precedence Diagram

Activity	Duration	Early Start	Late Start	Early Finish	Late Finish	Total Float
R						
D						
T						
Z						
F						
A						
H						
E						
S						
M						
B						
L						
V						
K						
C						
X						
P						

Table 3–7 Exercise 2—Data Table

3.

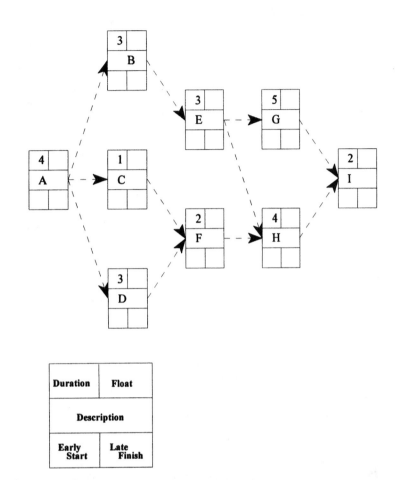

Figure 3–12   Exercise 3—Precedence Diagram

Activity	Duration	Early Start	Late Start	Early Finish	Late Finish	Total Float
A						
B						
C	.					
D						
E						
F						
G						
H						
I						

Table 3–8   Exercise 3—Data Table

4.

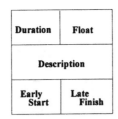

Duration	Float
Description	
Early Start	Late Finish

Figure 3–13 Exercise 4—Precedence Diagram

Activity	Duration	Early Start	Late Start	Early Finish	Late Finish	Total Float
A						
B						
C						
D						
E						
F						
G						
H						
I						
J						
K						

Table 3–9 Exercise 4—Data Table

5.

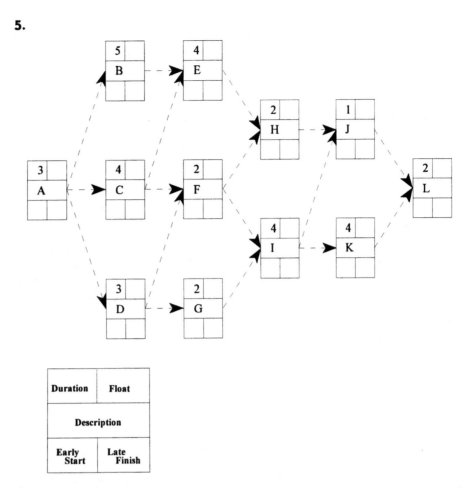

Figure 3–14   Exercise 5—Precedence Diagram

Activity	Duration	Early Start	Late Start	Early Finish	Late Finish	Total Float
A						
B						
C						
D						
E						
F						
G						
H						
I						
J						
K						
L						

Table 3–10   Exercise 5—Data Table

6.

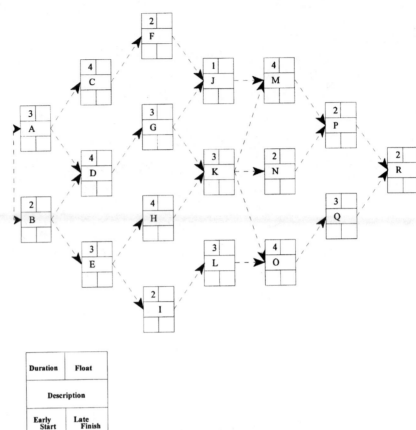

Figure 3–15   Exercise 6—Precedence Diagram

Activity	Duration	Early Start	Late Start	Early Finish	Late Finish	Total Float
A						
B						
C						
D						
E						
F						
G						
H						
I						
J						
K						
L						
M						
N						
O						
P						
Q						
R						

Table 3–11   Exercise 6—Data Table

7.

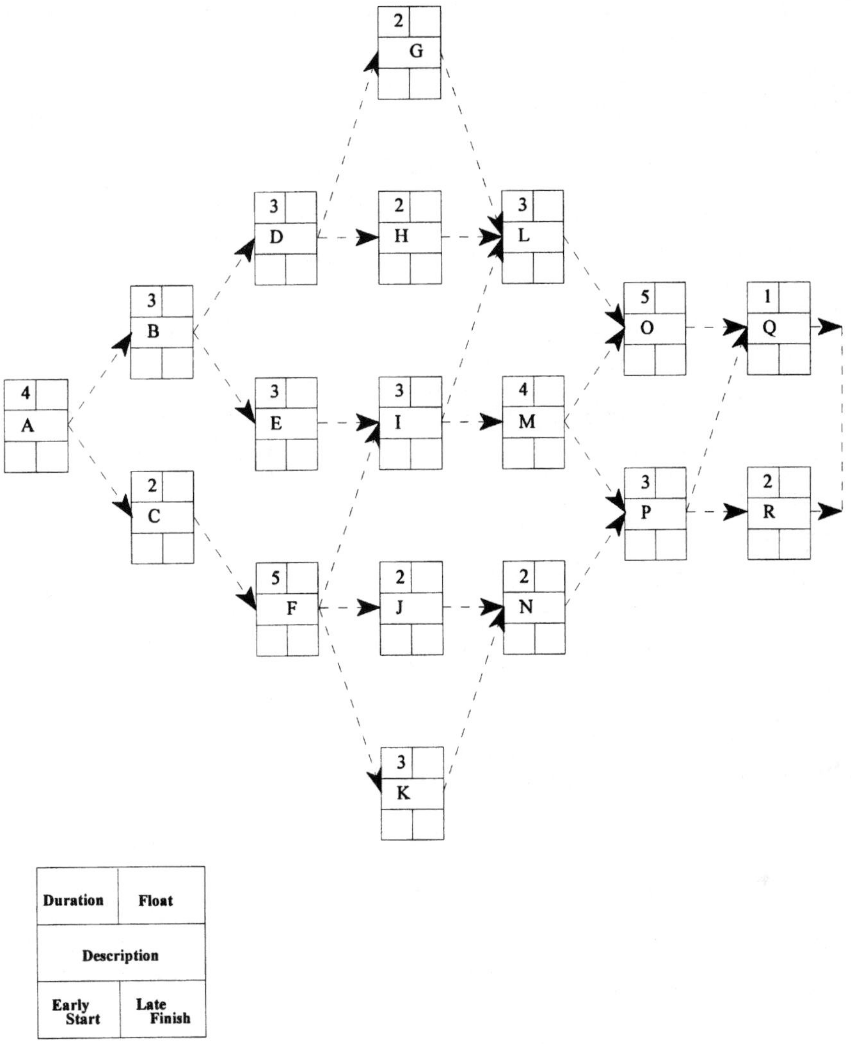

Figure 3–16   Exercise 7—Precedence Diagram

Activity	Duration	Early Start	Late Start	Early Finish	Late Finish	Total Float
A						
B						
C						
D						
E						
F						
G						
H						
I						
J						
K						
L						
M						
N						
O						
P						
Q						
R						

Table 3–12   Exercise 7—Data Table

8.

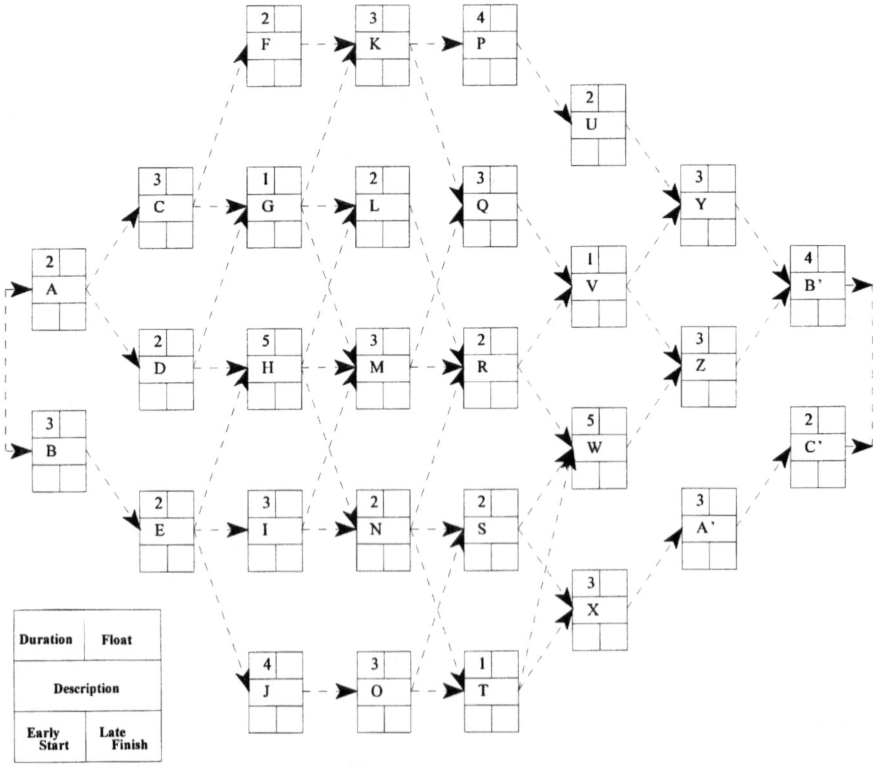

Figure 3–17   Exercise 8—Precedence Diagram

Activity	Duration	Early Start	Late Start	Early Finish	Late Finish	Total Float
A						
B						
C						
D						
E						
F						
G						
H						
I						
J						
K						
L						
M						
N						
O						
P						
Q						
R						
S						
T						
U						
V						
W						
X						
Y						
Z						
A'						
B'						
C'						

Table 3–13   Exercise 8—Data Table

4

Gantt Chart Creation

Objectives

Upon completion of this chapter, you should be able to:

- Add a project
- Add activities
- Define relationships
- Calculate the schedule
- Edit an activity
- Open a project
- Close a project
- Modify preferences

PROJECT MANAGER: ENTRANCE SCREEN

To start *P3e,* double click on the application icon (Figure 4–1) at the Windows Program Manager screen. The **Primavera Project Planner** dialog box (Figure 4–2) will appear. Using this dialog box, you can create a new schedule, open an existing schedule, open the last schedule used, or open global data (open empty or without a current project).

CREATE A NEW PROJECT SCHEDULE

To create a new schedule, click on the **Create New...** button of the **Primavera Project Planner** dialog box (Figure 4–2). The **Create a New Project** wizard (Figure 4–3) will appear. The **Create a New Project**

Figure 4–1 *P3e* Windows Icon

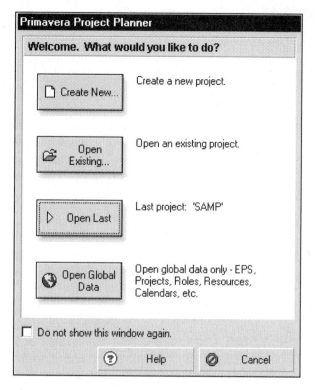

Figure 4-2   Primavera Project Planner Wizard

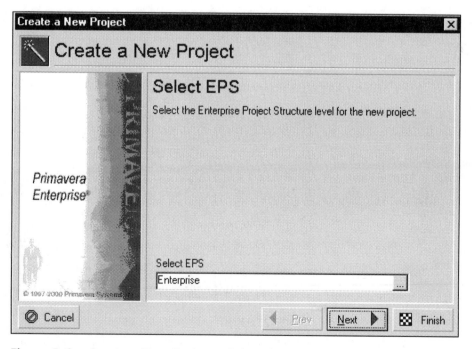

Figure 4-3   Create a New Project—Select EPS

wizard is a series of five screens used to gather information about the new project schedule being created.

Select EPS.   The first of the **Create a New Project** wizard screens (Figure 4–2) is the **Select EPS** screen (Figure 4–3). In *P3e*, the **Enterprise Project Structure** (EPS) forms the hierarchical structure of your database for projects. For the purposes of the Sample Schedule, **Enterprise** is selected as the EPS (Figure 4–3). The Sample Schedule, presently being created, will be used throughout this book for demonstration purposes.

Project Name.   The second of the **Create a New Project** wizard screens is the **Project Name** screen (Figure 4–4). The **Project ID** (the unique identifier for the project consisting of four alphanumeric characters) and **Project Name** are input. For the Sample Schedule about to be created, **SAMP** is the **Project ID**, and **Sample Schedule** is the **Project Name** selected.

Project Start and End Dates.   The third of the **Create a New Project** wizard screens is the **Project Start and End Dates** screen (Figure 4–5). Click the button at the right of the **Planned Start** window and a pop-up calendar will appear to use for selecting the planned start date for the

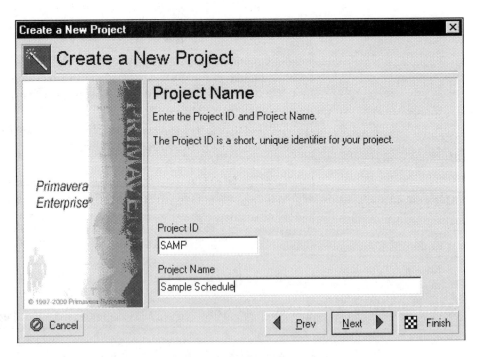

Figure 4–4   Create a New Project—Project ID and Name

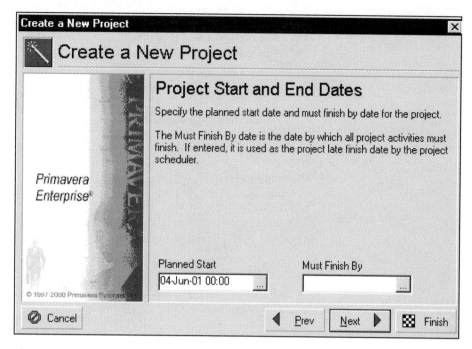

Figure 4–5   Create a New Project—Planned Start and End Dates

new schedule. June 4, 2001 is selected as the start date for the Sample Schedule. No finish date is selected, since, for the purposes of the Sample Schedule, we want *P3e* to calculate the project finish date.

Responsible Manager.   The fourth of the **Create a New Project** wizard screens is the **Responsible Manager** screen. The responsible manager is the top of the organizational breakdown structure (OBS) for the new project. For the purposes of the Sample Schedule, **Enterprise** was selected as the **Responsible Manager.**

Congratulations.   The fifth, and last, of the **Create a New Project** wizard screens is the **Congratulations** screen (Figure 4–6). By selecting the Finish button at the lower right hand corner of the **Create a New Project** wizard, the new project schedule is created.

NEW PROJECT ENTRANCE SCREEN

When the **Finish** button of the **Congratulations** screen (Figure 4–6) of the **Create a New Project** wizard (Figure 4–2) is clicked, *P3e* produces the **Home** screen of the newly created schedule (Figure 4–7). Note that the **Project ID (SAMP)** and **Project Name (Sample Schedule)** are now on the title bar at the top of the screen for the newly created schedule.

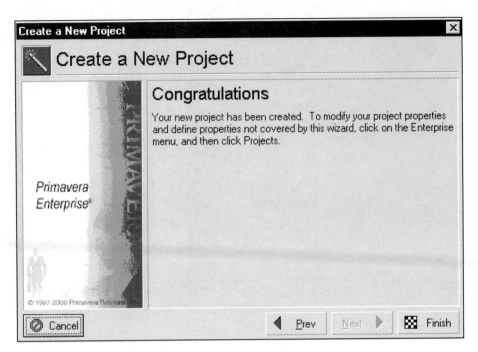

Figure 4–6   Create a New Project—Congratulations

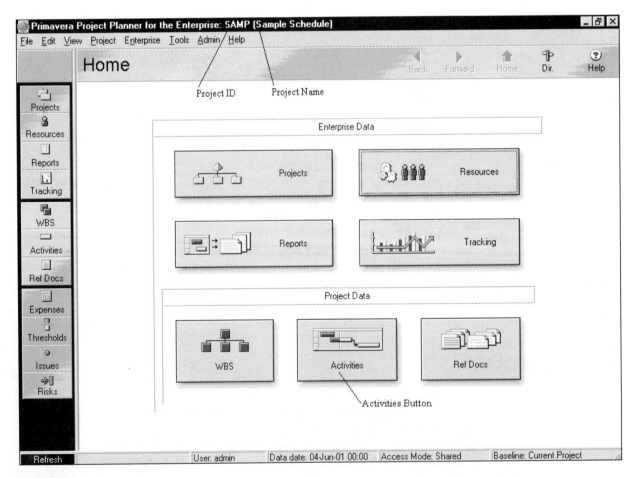

Figure 4–7   Home Screen

Enterprise Data.   The **Home** screen can be used to input, observe, or modify information that is applied across project lines. The four buttons under **Enterprise Data** are used to modify the database, which is applied to all projects.

Project Data.   The **Home** screen can also be used to input, observe, or modify the specific project information. The three buttons under **Project Data** are used to modify a specific project. The three buttons are **WBS** (Work Breakdown Structure), **Activities**, and **Ref Docs** (Reference Documents). To begin the process of inputting new information about the Sample Schedule, select the **Activities** button, under **Project Data** (Figure 4–7), and the **Activities** screen will be produced (Figure 4–8).

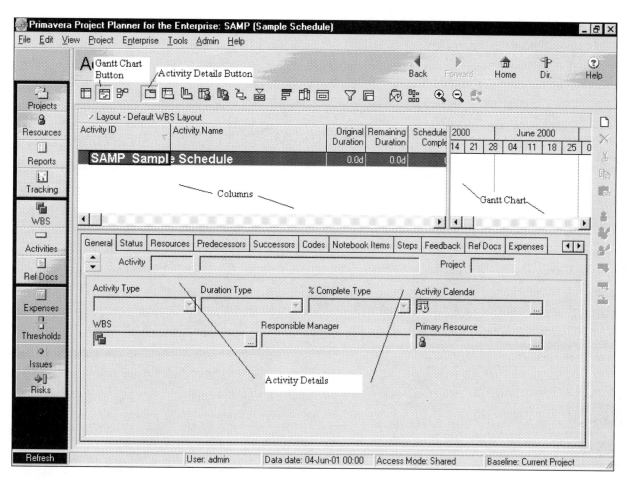

Figure 4–8   Activities Screen

Note the parts of the **Activities** screen shown in Figure 4–8 are columns, Gantt chart, and activity details.

For the purpose of clarity in producing the Sample Schedule, some of the toolbar features of the **Activities** screen will be turned off. Select the **View** main pull-down menu and then **Toolbars** (Figure 4–9). Note the on-screen locations of the **Toolbars** options in Figure 4–9. The **Toolbars** options that can be turned on and off are: **Navigation Bar, Navigation Bar Button Text, Directory, Directory Button Text, Activity Toolbar,** and the **Status Bar.** In Figure 4–10, all the **Toolbars** options have been unselected except the **Activity Toolbar.**

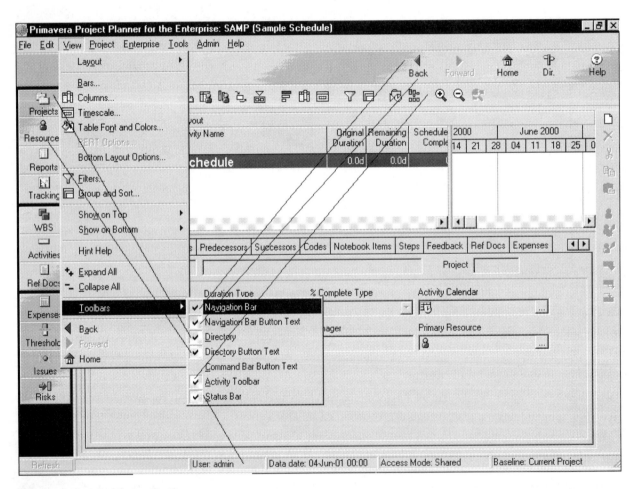

Figure 4–9   Toolbars Options

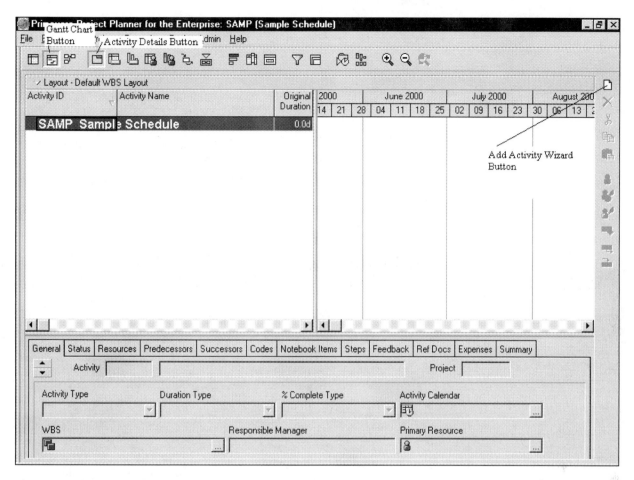

Figure 4–10   Activity Screen—Only Activity Toolbar Remains

ACTIVITIES

Figure 4–11 is the example logic diagram that is used throughout the rest of the book to explain the concepts in each chapter.

Before the process of adding activities is begun, one change needs to be made on the **Activity** screen. Note that under the **Original Duration** column (Figure 4–10), the format shown is 0.0d. To change the default format of the time units, select **User Preferences** from the **Edit** main pull-down menu. The **User Preferences** dialog box will appear (Figure 4–12). To modify the time unit format, select the **Time Units** tab. Then under **Format Durations,** click the down arrow of the **Units** window. The time unit options are **Hour, Day, Week,** and **Year. Day** is selected. Also, the **Decimals** option of **0** is selected. Now the **Original Duration** field is formatted to show the zero-day duration as 0d.

The steps in creating the Sample Schedule are to create the activities, input durations, and define relationships. *P3e* provides two ways for

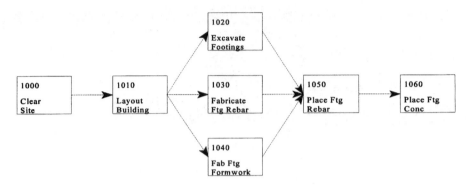

Figure 4–11   Sample Schedule

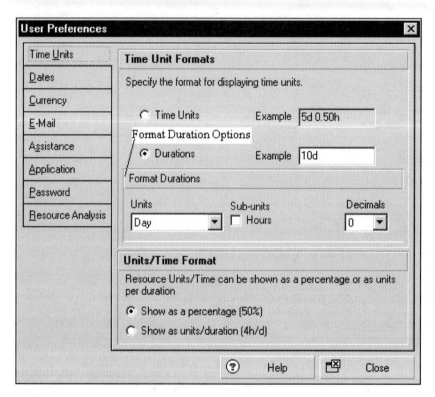

Figure 4–12   User Preferences Dialog Box

adding new activities. The first is by selecting the Add Activity Wizard from the Gantt Chart window (Figure 4–10). For the second method, and the method to be used in this book, select the **Add** option from the **Edit** main pull-down menu (Figure 4–13).

When the **Add** option is selected, a **New activity** is created on the Gantt Chart window (Figure 4–14). The **General** tab of the **Activity Details** window is selected to show the activity details on the bottom half of the Gantt Chart window. Compare Figure 4–14, with the new activity created, to Figure 4–10, without the new activity.

Figure 4–13   Edit Main Pull-Down Menu

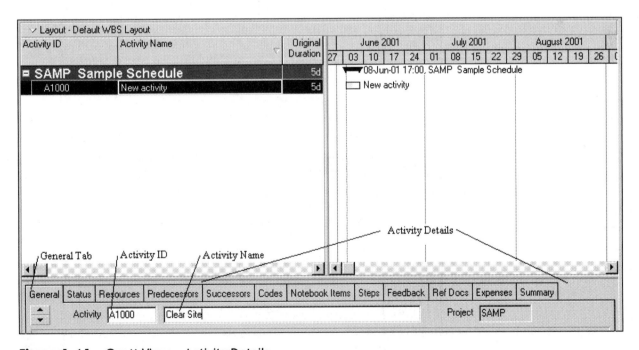

Figure 4–14   Gantt View—Activity Details

To define information about the added activity, the tabs and fields of **Activity Details** (Figure 4–14) will be used. The necessary fields for the new added activities will be discussed in the following section.

Activity.   An activity ID (identification) is a string of letters and/or numbers that uniquely identifies an activity. When entering activities, you can use the auto-sequencing feature, which numbers each new activity in increments of 10, starting with 1,000. To access the Activity Details, click on the **View** main pull-down menu then select **Show on Bottom** and then select **Activity Details** (Figure 4–15).

P3e automatically establishes the first activity with an activity ID of 1000. Each activity has a unique name, so *P3e* can use the unique designator to identify relationships used in establishing project duration, float, and so on. Since this is an alpha numeric field, it is necessary to use the same number of characters so that each activity is evaluated on a similar basis.

Activity ID.   When *P3e* assigns a new ID, it checks whether the ID already exists in the project. If it exists, *P3e* searches in increments of 10 until it finds an unused ID. If you delete an activity, that ID becomes available for a new activity. If you want *P3e* to increment IDs by a value other than 10, choose the **Enterprise** main pull-down menu, then **Projects,** and then the **Settings** tab. The **Settings** window will appear at the bottom of the screen (Figure 4–16). Type the value you want in the **Increment by** field. The **ID** field can be used to sort activities into ID categories within a project. To use this option, type in an **Activity ID Prefix** in the available window. Note that for the Sample Schedule, an **Activity ID Prefix** of **A** will be used. This means that the first activity will be **A1000.** After entering the ID, click on the **Activity Name** field.

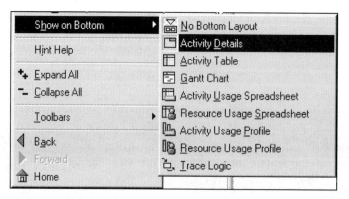

Figure 4–15   Activity Details Selection

| General | Dates | Notebook Items | Budget Log | Spending Plan | Budget Summary | Funding | Codes | Defaults | Resources | Settings |

Summarized Data

☐ Contains Summarized Data Only

Last Summarized On

Summarize to WBS Level

2 ▲▼

Baseline for Summarization

<Current Project> ▼

Auto-numbering Defaults

Settings Tab

Activity ID Prefix
A

Activity ID Suffix
1000

Increment
10

Project Settings

Character for separating code fields for the WBS tree   .

Critical activities have float less than or equal to   0d

Fiscal year begins on the 1st day of   January ▼

Figure 4–16   Projects View—Settings Tab

Activity Name. This alpha numeric field (Figure 4–14) is used to name or describe the activity. It is usually not a sort field within *P3e*. Use short, concise descriptions—usually two or three words in a noun + verb format. Note that in Figure 4–17, **Clear Site** is entered as the **Activity Name** for **Activity ID** of **A1000.** After entering the **Activity Name,** click on the **Original Duration** field of the Gantt chart.

Original Duration. If the scheduling unit is the day, then input the number of days for the activity duration. Note that in Figure 4–17, a two-day duration is input for the activity **A1000, Clear Site.** The appropriate field (A1000) from the **Original Duration** column was used to input the 2d original duration. Note also that there is no field available from the **General** tab of the **Activity Details** (Figure 4–18) for input of the original duration. If you want to input the original duration using the **Activity Details** fields, select the **Status** tab (Figure 4–18).

If you need to edit any activity information, simply double click in the wanted field while in the Gantt chart and type in the wanted changes.

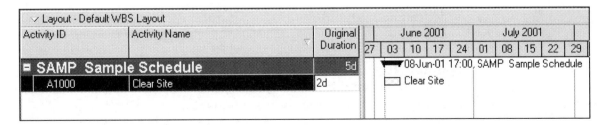

Figure 4–17   Sample Schedule—Added Activity

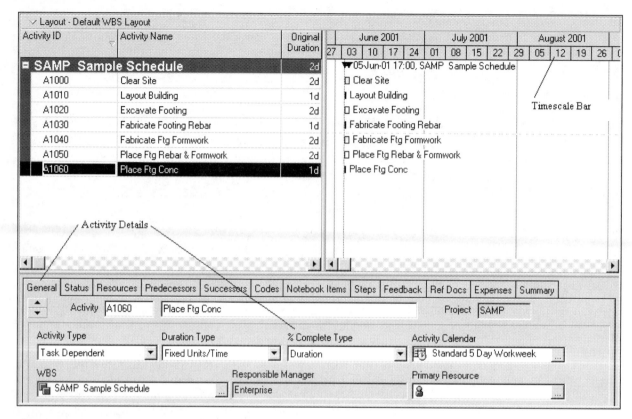

Figure 4–18 Activity Detail—General Tab

Activity Type. Select the **Activity Type** field (Figure 4–18). There are four options for **Activity Type.** They are:

Task Dependent. This option is the default, and the option used for activity **A1000, Clear Site.**

Resource Dependent. Use this option when the assigned resources' availability controls the activity schedule.

Level of Effort. Level of effort activities have durations that are driven by other activities.

Milestone. Milestones are zero-duration activities that mark significant project events.

Duration Type. Select the **Duration Type** field (Figure 4–18). Using the **Duration Type** field, *P3e* automatically synchronizes the duration, labor/nonlabor units, and resource units/time for activities so that the following equation is always true for each activity:

$$\text{Duration} = \text{Units} / (\text{Resource Units} / \text{Time}).$$

Since three variables are involved (duration, units, and units/time), when you change the value of one variable, *P3e* must alter the value of a second to balance the equation.

The three options for **Duration Type** are:

Fixed Units/Time.   **Fixed Units/Time** activities have resources with fixed productivity output per time unit. If an activity is designated as a **Duration Type** of **Fixed Units/Time,** then when you change the units/time assigned to the activity, *P3e* will automatically recalculate the activity duration based upon the new units/time input.

Fixed Duration.   This option is used for activity **A1000, Clear Site. Fixed Duration** activities will take the same amount of time no matter what resources are assigned to them. If the activity is designated as a **Duration Type** of **Fixed Duration,** when you change the duration assigned to the activity, *P3e* will automatically recalculate the activity units and units/time based upon the new duration.

Fixed Units.   **Fixed Units** activities require a fixed amount of work and will be finished faster if more resources are assigned. If the activity is designated as a **Duration Type** of **Fixed Units,** when you change the number of units or the units/time assigned to the activity, *P3e* will automatically recalculate the activity duration based upon the new number of units or units/time.

% Complete Type.   Choose the **% Complete Type** field (Figure 4–18). There are three options for **% Complete Type: Physical, Duration,** and **Units. Duration** is used for activity **A1000, Clear Site.**

Activity Calendar.   The Activity Calendar field from the exposed **General** tab of the **Activity Details** (see Figure 4–18) is used to define the calendar by which the activity will be scheduled. A base calendar defines when its activities can be worked on. Any activities to which you assign a particular base calendar can be worked on during the workdays in that base calendar, and they cannot be worked on during the nonworkdays in that base calendar. To create a base calendar, choose **Enterprise** and then **Calendars.**

The last three fields of **Activity Details (WBS, Responsible Manager, and Primary Resource)** will not be discussed.

Figure 4–18 shows all the new activities added for the Sample Schedule.

MODIFY TIMESCALE

Since the sample project is of short duration (Figure 4–18), the timescale of the bar chart needs to be modified to make the information easier to read.

Timescale Definition Dialog Box

Timescale Density. The timescale density (the amount of time units appearing on the screen) can be controlled by left clicking and dragging the lower portion of the timescale bar. The **Timescale** dialog box can also be used for making timescale density changes.

The **Timescale** dialog box (Figure 4–19) can be accessed in one of two ways. The first is to click on the **View** main pull-down menu, and then click on the **Timescale** option. The other method is to right click anywhere on the shaded timescale bar itself.

Modifications that may be made using the **Timescale** dialog box are:

> **From Start Date.** The value of **PS** is shown in Figure 4–19. This indicates that the on-screen bar chart starts at the **Planned Start** given when the schedule was created (Figure 4–5).

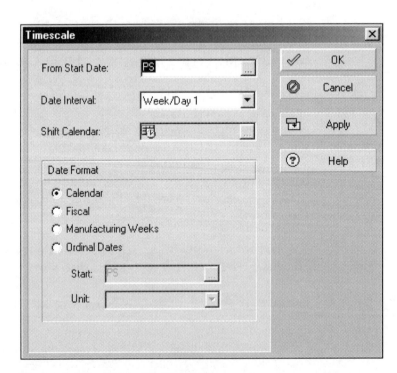

Figure 4–19 Timescale Dialog Box

The **From Start Date** options are:

<u>P</u>S: **Earliest Project Start**

P<u>F</u>:   **Latest Project Finish**

<u>D</u>D: **Earliest Data Date**

<u>C</u>D: **Current Date**

C<u>W</u>:   **Current Week**

C<u>M</u>:   **Current Month**

     <u>C</u>ustom Date

Date Interval (Figure 4–19).   The timescale option shown in Figure 4–18 is weeks. When Figure 4–20 was executed, days were selected. Notice the difference shown on the timescale bar. Figure 4–18 shows only the Mondays (beginning of the week) that are available. Figure 4–20 shows all days. For Figure 4–20, the **Week/Day 1** option was chosen. The week scale is at the top of the timescale (Jun 03, Jun 10, Jun 17, and Jun 24). The day scale is at the bottom of the timescale (S,M,T,W,T,F,S). The other options for the **Date Interval** field are: **Year/Quarter, Year/Month, Quarter/Month, Month/Week, Week/Day 1, Week/Day 2, Day/Shift,** and **Day/Hour.** To change the time frame, use the down arrow button.

Shift Calendar (Figure 4–19).   This option is available only if the **Day/Shift** option is chosen in the **Date Interval** field. Here the corresponding shift is chosen.

Date Format (Figure 4–19).   The **Date Format** options are **Calendar, Fiscal, Manufacturing Weeks,** and **Ordinal Dates.** For the Sample Schedule, the **Calendar** option is selected.

Usually it is easier to think of your schedule in terms of calendar days. But sometimes, particularly for the beginning scheduler, it is simpler to see the ordinal (or numerical) workdays. The manual schedules used in Chapter 3 to calculate the examples all used ordinal dates. By clicking on the **Ordinal dates** selection box, the timescale bar is changed from calendar to ordinal days.

RELATIONSHIPS

All the activities in the Sample Schedule start on the same date, the first day of the project (see Figure 4–20). This is because no relationships have

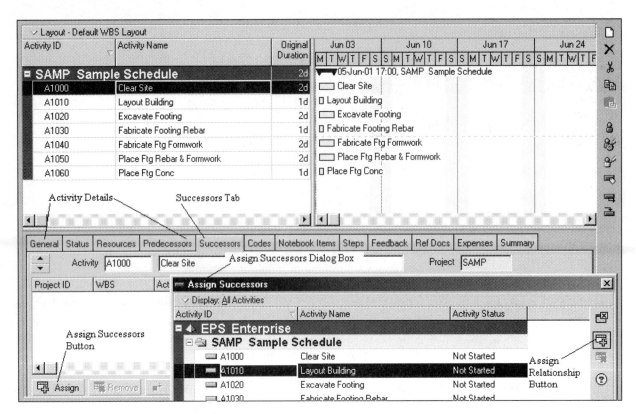

Figure 4–20 Assign Successors Dialog Box

been established between the activities yet. After establishing the activity ID and giving the activity a name and an original duration, the next step is to establish relationships among the activities. The procedure for establishing relationships is as follows:

Successor/Predecessor. An activity relationship can be either a predecessor or a successor relationship. If Activity A must occur before B, A is a predecessor to B and B is a successor to A. Either of these relations can be identified to *P3e*. If one is identified, the program automatically assumes the other. Click on the activity to which relationships are to be established. In Figure 4–20, Activity A1000, Clear Site, is selected. Click the right mouse button and select **Activity Details** and then the **Successors** tab (see Figure 4–20). To define the new relationship, click the **Assign** successor button, at the lower right-hand corner of the **Successors** tab of the **Activity Details** window. The **Assign Successors** dialog box will appear (Figure 4–20). Using Figure 4–11, Activity A1010 should be the successor to Activity A1000. With Activity A1000 highlighted, Activity A1010 is selected in the **Assign Successors** dialog box. Then the **Assign** relationship button is clicked at the right of the **Assign Successors** dialog box (Figure 4–20). Now look at the **Successors** tab of the **Activity Details** window (Figure 4–21). Activity A1010 is shown as a successor to Activity A1000. The relationship line now appears on the screen in the Gantt chart area showing the established relationship, if relationship lines

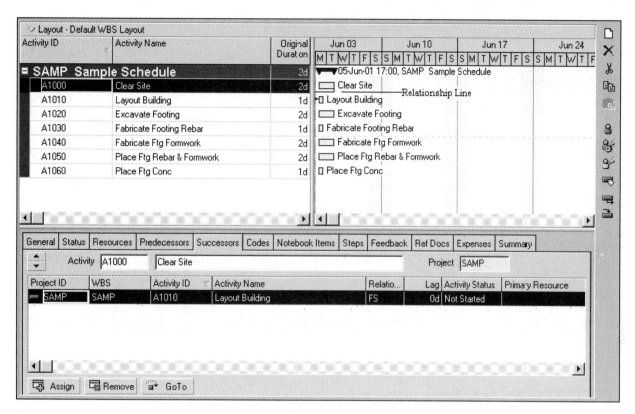

Figure 4–21 Activity Details—Successors Tab

are turned on for viewing. The reason that Layout Building (Activity A1010) does not physically come after Clear Site (Activity A1000) on the Gantt chart is that the changes have not been "Scheduled" yet.

Relationship. When Activity A1010 is selected as a successor to Activity A1000, the default relationship is **FS - Finish to start** relationship (Figure 4–21 and Figure 2–1). Activity A1000 must be finished before Activity A1010 can start. Click in the **Relationship** field and then click on the down arrow on the tool bar to see a pull-down menu of other relationship types. There are four relationship types—the other three are **SS, FF,** and **SF.** An **SS - Start to start** relationship means the successor activity can start at the same time or later than the predecessor activity. An **FF - Finish to finish** means that the successor activity can finish at the same time as or later than the predecessor. An **SF - Start to finish** means that the predecessor activity must start before the successor activity can finish. Clicking on one of the other options changes the default **FS** relationship. Multiple successors to an activity are shown in Figure 4–22. This can be done by simply repeating the preceding steps.

Lag. Lag is a delay or offset time in a relationship or resource assignment. If you specify two days of lag in a finish to start (FS) relationship, the successor activity starts two days after the predecessor finishes. In Figure 4–22, Activity A1030 will start 1 day after the finish of Activity

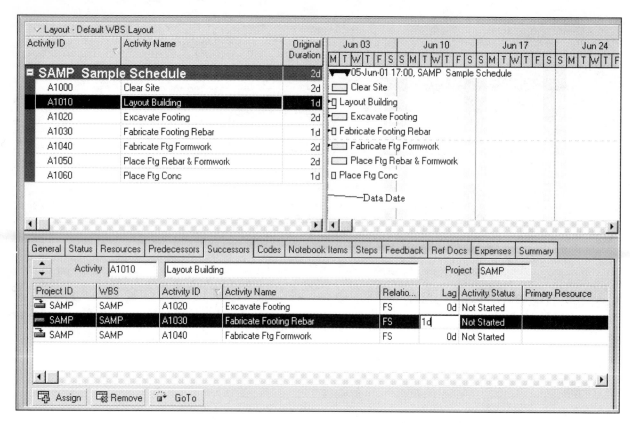

Figure 4–22 Activity Details—Successors Assigned

A1010. A lag of 1 day is imposed on the relationship. Click in the **Lag** field of the activity that should have lag; then click on the number field. For example, lag time is given to allow elevated concrete to cure to a certain level before a formwork wrecking/reshoring activity could begin. A negative lag may also be introduced and in this case will be called "overlap."

Dependencies or relationships can be displayed within *P3e.* Click the **View** main pull-down menu, select **Bars,** and then select the **Options** button. Click the selection to display relationships on the Gantt chart screen.

For another way to establish new relationships, place the cursor at the beginning or end of an activity, which causes the relationship line symbol to appear. Hold the left mouse button down and drag this symbol from activity to activity to establish the same relationships as using the methods just described. This method for establishing relationships is discussed in detail in the chapter on PERT diagramming later in the book. To correct an error made in inputting activity relationships, see the Edit Activities section.

Table 4–1 lists the activities, the successor activity IDs, type relationships, and the lags input for the Sample Schedule. These are calculated in the next section.

Activity	Successor Activity ID	Relationship	Lag
1000	1010	FS	0
1010	1020	FS	0
	1030	FS	0
	1040	FS	0
1020	1050	FS	0
1030	1050	FS	0
1040	1050	FS	0
1050	1060	FS	0
1060			

Table 4–1   Sample Schedule Activity Relationships

CALCULATE THE SCHEDULE

The Schedule Dialog Box

To obtain the **Schedule** dialog box, there are three methods.

Method 1.   Click on the **Tools** main pull-down menu. Then click on **Schedule**.

Method 2.   Press the **F9** function key on the keyboard.

Method 3.   Click on the Schedule Toolbar button and the Schedule Dialog box will appear.

The **Schedule** dialog box is displayed (Figure 4–23). The **Current Data Date** establishes the blue vertical line (Figure 4–22) that reflects the date that will be the basis of the schedule, which can be either the planned start date or an interim update date. The **Current Data Date** of 04-Jun-01 appears in Figure 4–23. This date appears since we input this date (Figure 4–5) as the **Planned Start** date of the sample schedule.

Next, click on the **Schedule** button (Figure 4–23) and the schedule is calculated. Figure 4–24 is the result of the calculated schedule for the Sample Schedule.

EDIT ACTIVITIES

Activity descriptions, original durations, relationships, or other input data can easily be changed or added to an existing activity. Click on the

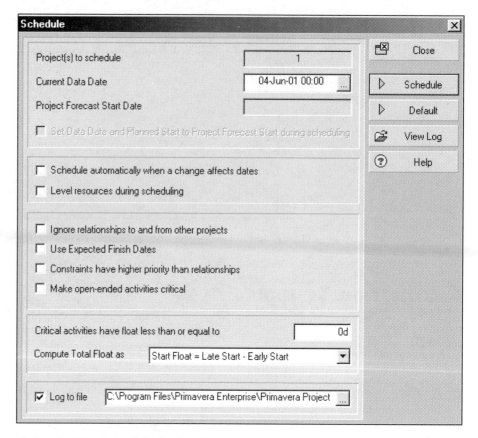

Figure 4–23   Schedule Dialog Box

Activity ID	Activity Name	Original Duration	
SAMP Sample Schedule		**8d**	13-Jun-01 17:00, SAMP Sample Schedule
A1000	Clear Site	2d	Clear Site
A1010	Layout Building	1d	Layout Building
A1020	Excavate Footing	2d	Excavate Footing
A1030	Fabricate Footing Rebar	1d	Fabricate Footing Rebar
A1040	Fabricate Ftg Formwork	2d	Fabricate Ftg Formwork
A1050	Place Ftg Rebar & Formwork	2d	Place Ftg Rebar & Formwork
A1060	Place Ftg Conc	1d	Place Ftg Conc

Figure 4–24   Sample Schedule—Schedule Calculated

activity to be modified. Turn the **Activity Details** window on (if it is off) by clicking on **View** from the main pull-down menu. Then click on the **Show on Bottom,** and then choose **Activity Details.**

Any of the fields for the highlighted activity may be modified. While still in this edit mode with the **Activity Details** turned on, other activities can be edited by clicking on them.

P3e has a number of other handy editing tools available under the **Edit** main pull-down menu (Figure 4–25). By selecting the **Find** option, the

Find dialog box is brought up (Figure 4–26). This option is used to find a key word or entry in a schedule with many activities. Note that Footing was entered in the **Find wha**t field, and after clicking the **Find Next** button, *P3e* found Activity A1030, Fabricate Footing Rebar. Another editing tool found on the **Edit** main pull-down menu (Figure 4–25) is the

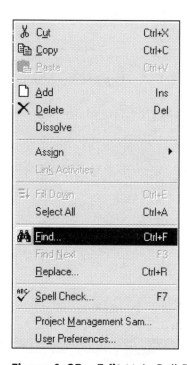

Figure 4–25   Edit Main Pull-Down Menu

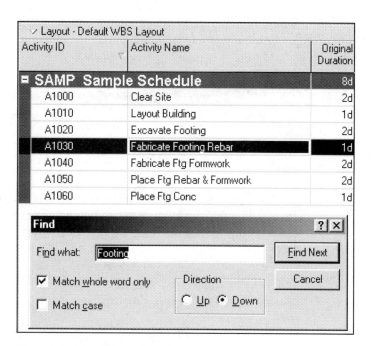

Figure 4–26   Find Dialog Box

Spell Check option. By selecting the **Spell Check** option, the **Spell Check** dialog box is brought up (Figure 4–27). Note, from Figure 4–27, that *P3e* found that rebar was either misspelled or not in the spelling dictionary. Since rebar is a common abbreviation for reinforcing, by clicking the **Add** button, the term rebar can be added to the dictionary.

CLOSE A PROJECT

P3e provides three methods to close an existing project:

Method 1.   Click on the **File** main pull-down menu. Click on **Close: All.** This closes the particular schedule but does not shut down *P3e.*

Method 2.   Click on the **P3e Project Manager** pull-down menu. Click on **Close.** This not only closes the particular schedule, it also shuts down *P3e.*

Method 3.   Simultaneously press the **Alt** and the **F4** keys. This closes the particular schedule and shuts down *P3e.*

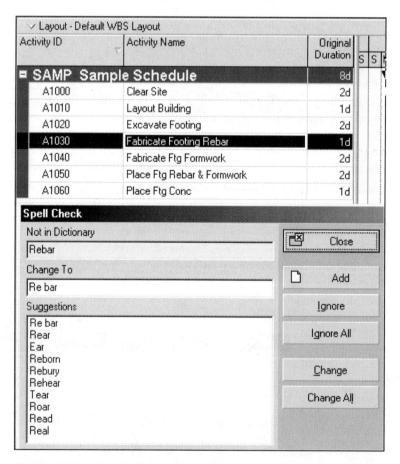

Figure 4–27   Spell Check Dialog Box

OPEN A PROJECT

To open an existing schedule rather than creating a new schedule as described in this chapter, click on the **File** main pull-down menu and select **Open**. The **Open Project** dialog box of existing projects appears (Figure 4–28). *P3e* is not file-based. It reads from a database. There are no file directories or drives.

PROJECT MAINTENANCE

P3e stores all project and resource/role data in central databases. You can import and export information to and from these databases using external files. The ability to import information from another database can be a real time saver for the scheduler. You can also use external files to archive projects no longer being currently used or to create a backup of the databases.

Figure 4–28   Open Project Dialog Box

From the **File** main pull-down menu, there are four options available for project maintenence. They are:

Import and Export.   The first two of the file maintenance options from the **File** main pull-down menu, **Import** and **Export** (Figure 4–29), are used to copy projects and share information among databases. There are two types of data that can be imported or exported. The first type of data that can be shared is resource and role information. Resource and role information can be imported and exported using .XER file format (the Primavera proprietary file format). The second type of data that can be imported or exported are projects. The export function is used to copy an existing project schedule and export the file to another location. The import function is used to bring a project schedule in from an external location and place it in the *P3e* database.

P3e stores all project data in a central database. *P3e* supports three types of file formats when importing and exporting project information. The first supported type of file format is .XER, the Primavera propriety file format. The second supported type of file format is .MPX, the *Microsoft Project* Exchange file format. The third supported type of file format is the P3 3.0 format, which enables you to share information with *Primavera Project Planner* version 3.0.

Check-in and Check-out.   These features are used to check-in or to check-out a project and to keep track of which projects have been checked out. This is a very handy feature when projects are being updated.

New...	Ctrl+N
Open...	Ctrl+O
Close All	Ctrl+W
Login as a Different User...	
Page Setup...	
Print Preview	
Print...	Ctrl+P
Import...	
Export...	
Check In...	
Check Out...	
Import from Expedition...	
Commit Changes	
Refresh Data	F5
Exit	

Figure 4–29   File Main Pull-Down Menu

TABLE FONT AND COLORS

Select **Table Fo<u>n</u>t and Colors** from the <u>V</u>iew main pull-down menu. The **Table Font and Colors** dialog box will appear (Figure 4–30). This dialog box is used to change the appearance of the on-screen table text, color, and row height. The **Table Font and Colors** dialog box options are:

1. **<u>F</u>ont.**   Click on the **Font** button to see the **Font** dialog box (Figure 4–31). Here you can choose a default font (typeface) for the contents of the project window. If a project is open, you are establishing the default font for the project. If no projects are open, you are establishing the default font for all future projects.

2. **Color.**   Click on the **Color** button of the **Table Font and Colors** dialog box (Figure 4–30) to bring up the **Color** dialog box (Figure 4–32). This dialog box is used to change the appearance of the on-screen display's background color.

3. **Row Height.**   Click on the **Row Height** window of the **Table Font and Colors** dialog box (Figure 4–30) to change the row height of the on-screen display. Changing the row height is a handy feature for adjusting the amount of information displayed on the screen, and it is very useful in customizing hard-copy prints to get just the number of activities on a page that you want to see.

4. **Apply.**   Click on the **Apply** button of the **Table Font and Colors** dialog box (Figure 4–30) to apply your changes to the on-screen display without closing the dialog box.

5. **OK.**   Click on the **OK** button of the **Table Font and Colors** dialog box (Figure 4–30) to save your changes to the on-screen display and close the dialog box.

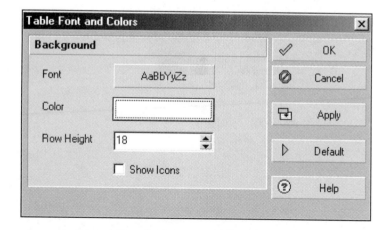

Figure 4–30   Table Font and Colors Dialog Box

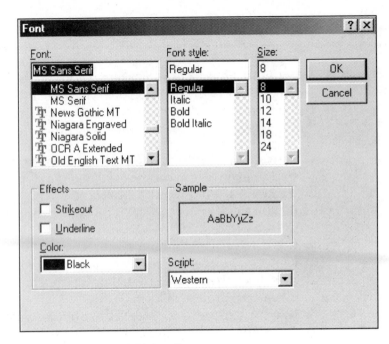

Figure 4–31 **Font** Dialog Box

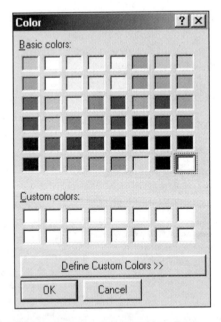

Figure 4–32 **Color** Dialog Box

WIZARDS

Wizards are a great way to speed up your work. They quickly guide you through repetitive tasks, doing most of the work for you. *P3e* includes the following wizards:

- Create projects
- Set administrative preferences
- Add activities
- Add resources
- Create reports
- Import projects
- Export projects
- Check-in projects
- Check-out projects

You can choose to launch the **New Activity** or **New Resource** wizards automatically when you add a new activity or resource. The other wizards are available when you select the applicable menu option.

PREFERENCES

Administrative Preferences

P3e allows you or your organization to define a number of schedule variables that will apply to all projects in the database, not just the project being currently worked upon. Whereas all *P3e* users can view these settings, special security privileges can be assigned to users to limit editing. One of the ways *P3e* allows you to define schedule variables is the use of the **Admin Preferences** dialog box (Figure 4–33). To get to the **Admin Preferences** dialog box, select **Admin Preferences** from the **Admin** main pull-down menu. As can be seen from Figure 4–33, the tabs available under the **Admin Preferences** dialog box are:

- **General**
- **Timesheets**
- **Progress Reporter**
- **Data Limits**
- **ID Lengths**
- **Time Periods**
- **Earned Value**
- **Reports**
- **Currency**
- **Options**
- **User Fields**

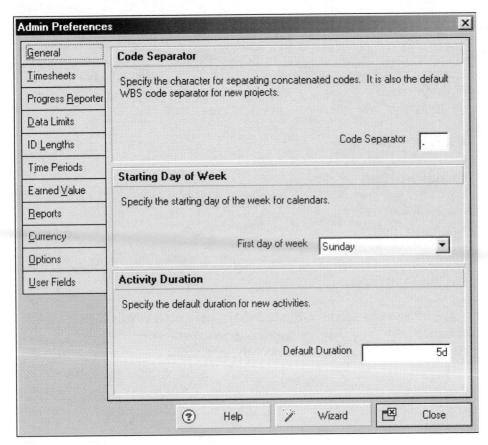

Figure 4–33 **Admin Preferences** Dialog Box—**General** Tab

It is not necessary to look at all these tabs and fields at this point in the development of the Sample Schedule. We will discuss some of the schedule variables available from the **Admin Preferences** dialog box that are of particular interest now.

General Tab. The **General** tab of the **Admin Preferences** dialog box is shown in Figure 4–33. The default value for **Starting Day of the Week** is defined. This is the starting day of the week for the project to be used by all defined calendars. The default value for **Activity Duration** is also defined. The value defined in this field will set the default duration for all new activities.

ID Lengths Tab. The ID **Lengths** tab of the **Admin Preferences** dialog box is shown in Figure 4–34. The **ID Lengths** fields are used to define the default width of ID fields available within *P3e*. For example, the **Activity ID maximum characters** field is set at 20 characters in Figure 4–34. Observe the activity ID field in Figure 4–34. If for some reason, a wider field is needed to define the activity ID, this field could be used to make the modification.

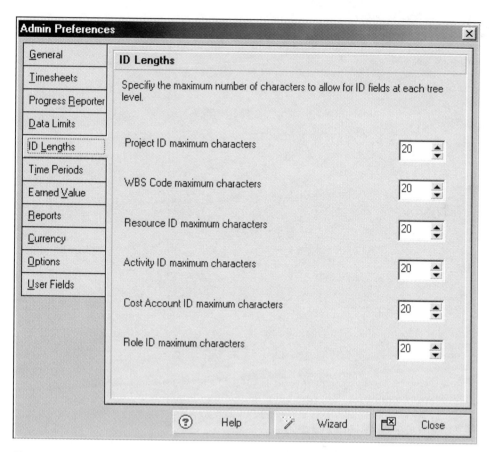

Figure 4–34   Admin Preferences Dialog Box—**ID Lengths** Tab

Time Periods Tab.   The **Time Periods** tab of the **Admin Preferences** dialog box is shown in Figure 4–35. The **Hours per Time Period** fields are used to establish the number of hours per work periods to be used by internal *P3e* calculations for items such as activity durations and resource usage. The **Time Period Abbreviation** field is used to establish the abbreviations that will appear on-screen for time units.

User Preferences

The **User Preferences** dialog box (Figure 4–36) is used to specify *P3e* settings and preferences for a specific user. To get to the **User Preferences** dialog box, select **User Preferences** from the **Edit** main pull-down menu. As can be seen from Figure 4–36, the tabs available under the **User Preferences** dialog box are:

- **Time Units**
- **Dates**
- **Currency**
- **E-Mail**

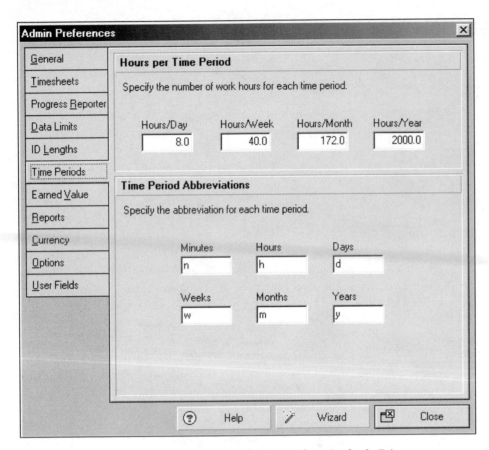

Figure 4–35 **Admin Preferences** Dialog Box—**Time Periods** Tab

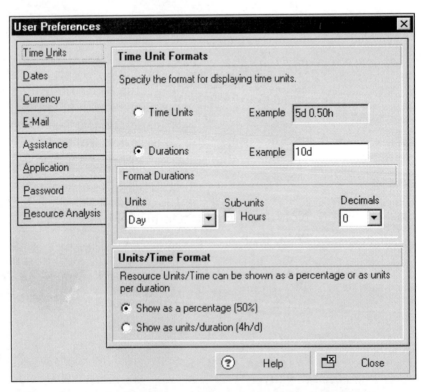

Figure 4–36 **User Preferences** Dialog Box—**Time Units** Tab

- A̲ssistance
- A̲pplication
- P̲assword
- R̲esource Analysis

It is not necessary to look at all the tabs and fields available at this point in the development of the Sample Schedule. Some of the schedule variables available from the **User Preferences** dialog box that are of particular interest now will be discussed.

Time Units Tab.   The **Time U̲nits** tab of the **User Preferences** dialog box is shown in Figure 4–36 and Figure 4–12. This tab enables you to define the displayed time units of the on-screen timescale. The time unit selected affects how *P3e* displays activity durations, tracking layout, resource prices, resource availability, and value formulas. From Figure 4–36, the Day was selected as the time **Units** option for the Sample Schedule. The time **Units** possible options are: Hour, Day, Week, Month, and Year. The **Sub-units** checkbox can be used to convert from Day (or major unit) to a Day/Hour format. The **Durations Example** of 10d shows how the activity duration values will appear. If the **Sub-units** checkbox had been checked, the **Durations Example** would show 10d 1h. The **Decimals** options are 0, 1, or 2 decimal places to be shown by the time unit.

Dates Tab.   The **D̲ates** tab of the **User Preferences** dialog box is shown in Figure 4–37. The **D̲ates** tab enables you to specify how the dates will be displayed in *P3e*. The **Date Format** fields allow you to specify the order of the date information. The **Date Format** options are:

- **M̲onth, Day, Year**
- **Day, Mont̲h, Year**
- **Y̲ear, Month, Day**

For the Sample Schedule, **Day, Mont̲h, Year** is selected. Note that, from the **Sample** field, this format will be displayed as 25-Feb-01.

The **Time** field of the **D̲ates** tab enables you to select **12 hour, 24 hour,** or **Don't sho̲w time** options. For the Sample Schedule, **24 hour** is selected. Note that, from the **Sample** field, this format will be displayed as 25-Feb-01 00:00.

Under the **Options** field, the **4-digit year** checkbox is left unchecked. This means that from the **Sample** field, the date appears as 25-Feb-01 instead of 25-Feb-2001. The **Month n̲ame** checkbox is checked. This means that the date appears as 25-Feb-01 instead of 25-02-2001. The **L̲eading zeros** checkbox is checked. This means that the date appears as 4-Feb-01 instead of 04-Feb-01. The **S̲eparator** field has the "–" character

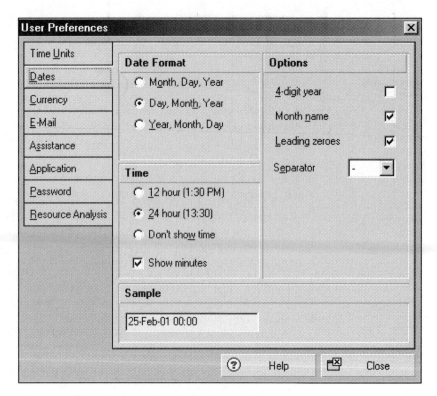

Figure 4–37   User Preferences Dialog Box—**Dates** Tab

selected as the separator between the components of the date. If the "." had been chosen instead, the date would appear as 25.Feb.01 instead of 25-Feb-01.

EXAMPLE PROBLEM

Table 4–2 is a tabular list of 28 activities (activity ID, description, duration, and successor relationships) for a house put together as an example for student use (see the wood-framed house drawings in the Appendix). Figure 4–38 is the on-screen Gantt chart of the house based on the list provided in Table 4–2.

Activity ID	Act Description	Duration (Days)	Successors
A1000	Clear Site	2	A1010
A1010	Building Layout	1	A1020
A1020	Form/Pour Footings	3	A1030
A1030	Pier Masonry	2	A1040
A1040	Wood Floor System	4	A1050
A1050	Rough Framing Walls	6	A1060
A1060	Rough Framing Roof	4	A1080
A1070	Doors and Windows	4	A1090, A1150
A1080	Ext Wall Board	2	A1070, A1100, A1110
A1090	Ext Wall Insulation	1	A1160
A1100	Rough Plumbing	4	A1130
A1110	Rough HVAC	3	A1120
A1120	Rough Elect	3	A1090
A1130	Shingles	3	A1090, A1150, A1260
A1140	Ext Siding	3	A1190
A1150	Ext Finish Carpentry	2	A1140
A1160	Hang Drywall	4	A1170
A1170	Finish Drywall	4	A1180, A1230, A1240
A1180	Cabinets	2	A1200, A1220
A1190	Ext Paint	3	A1210
A1200	Int Finish Carpentry	4	A1250
A1210	Int Paint	3	A1210
A1220	Finish Plumbing	2	A1250
A1230	Finish HVAC	3	A1210
A1240	Finish Elect	2	A1210
A1250	Flooring	3	A1270
A1260	Grading & Landscaping	4	A1270
A1270	Punch List	2	

Table 4–2   Activity List with Durations and Relationships—Wood-Framed House

EXERCISES

1. Create an on-screen *P3e* Gantt chart for Exercise 1 in Chapter 3.
2. Create an on-screen *P3e* Gantt chart for Exercise 2 in Chapter 3.
3. Create an on-screen *P3e* Gantt chart for Exercise 3 in Chapter 3.
4. Create an on-screen *P3e* Gantt chart for Exercise 4 in Chapter 3.
5. Create an on-screen *P3e* Gantt chart for Exercise 5 in Chapter 3.
6. Create an on-screen *P3e* Gantt chart for Exercise 6 in Chapter 3.
7. Create an on-screen *P3e* Gantt chart for Exercise 7 in Chapter 3.

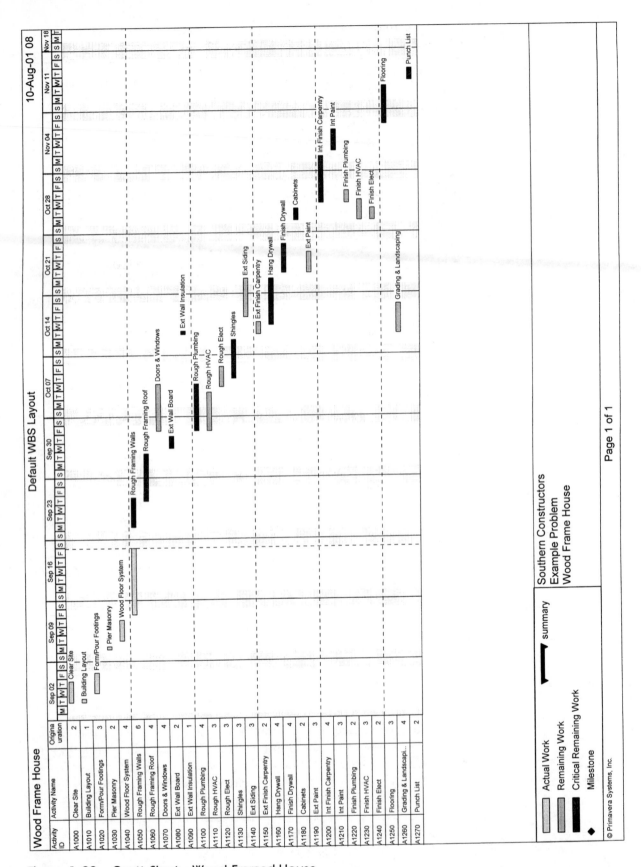

Figure 4–38   Gantt Chart—Wood-Framed House

8. Create an on-screen *P3e* Gantt chart for Exercise 8 in Chapter 3.

9. **Small Commercial Concrete Block Building—On-Screen Gantt Chart**

Prepare an on-screen Gantt chart for the small commercial concrete block building located in the Appendix. This exercise should include the following steps:

A. Prepare a list of activity descriptions (minimum of 60 activities).

B. Establish activity durations.

C. Establish activity relationships.

D. Create the on-screen *P3e* Gantt chart.

10. **Large Commercial Building—On-Screen Gantt Chart**

Prepare an on-screen Gantt chart for the large commercial building located in the Appendix. This exercise should include the following steps:

A. Prepare a list of activity descriptions (minimum of 150 activities).

B. Establish activity durations.

C. Establish activity relationships.

D. Create the on-screen *P3e* Gantt chart.

5

Gantt Chart Format

Objectives

Upon completion of this chapter, using *P3e*, you should be able to:

- Utilize layouts
- Modify columns
- Modify bars
- Show relationships
- Modify sight lines
- Modify row height
- Use activity codes
- Group and sort activities
- Summarize activities
- Filter activities

One of the major advantages of using the computer for scheduling construction project management applications is the ease of modifying on-screen and hard-copy printouts of the schedule to communicate information in the clearest way possible. This includes changing the appearance and content of columns, bars, row heights, sight lines, etc.; organizing the information using activity codes; and selecting particular activities included in the schedule using filtering. All this flexibility assures that *P3e* can produce a schedule in the format and with the information to fit the multiple scheduling needs of a complex project.

LAYOUTS

The appearance of the on-screen schedule depends on its layout. A *layout* is a stored format of your schedule that keeps visual elements such as the format of activity bars, row heights, site lines, columns, the organization of activities, and screen colors. To select layouts, click on the **View** main pull-down menu (Figure 5–1).

The options under **Layout** are: **Open, Save,** and **Save As.** These layout options provide you with the ability to change the visual appearance of

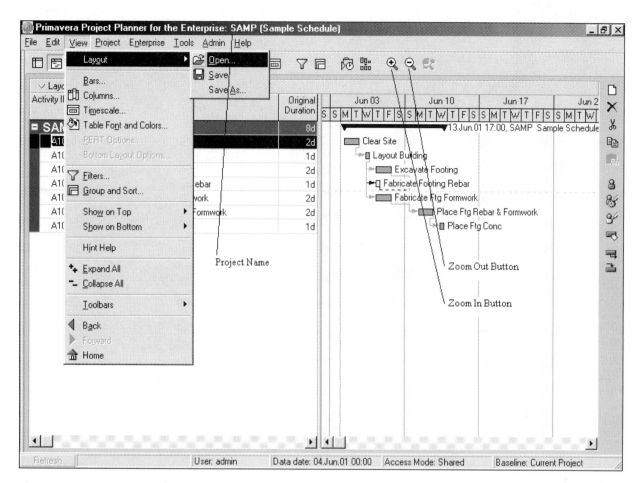

Figure 5–1   View Main Pull-Down Menu

screens and to keep these changes intact when the program is restarted. Also, multiple screens (and therefore hard copies) of different layout formats can be saved and reused.

Save.   *P3e* automatically saves changes to activity data such as adding or editing activities. To save changes to the appearance of a layout, such as changing the column data, modifying bar sizes and colors, or displaying the activity form, use this **Save** option, under **Layout,** from the **View** main pull-down menu.

Save As.   The **Save As** selection from the **Layout** options opens the **Save Layout As** dialog box (Figure 5–2). When you first create a layout, use the **Save As** command to assign a new Layout Name (Sample Schedule Layout in Figure 5–2).

Open.   The **Open** selection from the **Layout** options opens the **Open Layout** dialog box (Figure 5–3). Here you can open an existing layout. *P3e* opens only one layout at a time for each project. In Figure 5–3, the

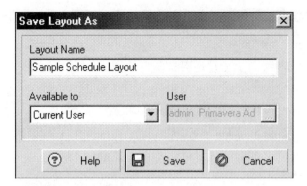

Figure 5–2   Save Layout As Dialog Box

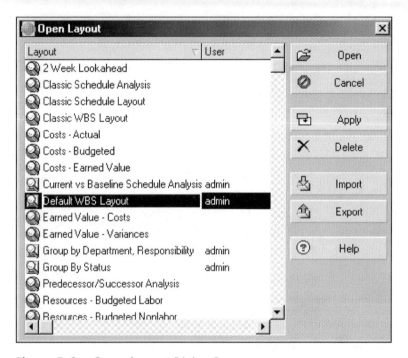

Figure 5–3   Open Layout Dialog Box

Default WBS Layout has been selected. The fields available within the **Open Layout** dialog box to help in selecting a layout are:

Layout.   The **Layout** field lists the name of each available layout.

User.   The **User** field lists the login name of the user who created the layout.

Apply.   The **Apply** button is used to preview the layout without closing the **Open Layout** dialog box.

Delete.   The **Delete** button is used to delete the selected layout from the **Open Layout** dialog box.

Import.   You can import a layout from a different database or user to the current one. Layouts are not project specific. They are user and database specific. Also, Import can be used to quickly create a new layout or to transfer a similar one; you can then rename it and modify it. The **Import** layout command opens the **Import** dialog box (Figure 5–4). Select the file that contains the layout you want to import and then click the **Open** button. *P3e's* layout files have a .PLF (Primavera Layout File) format. *P3e* appends a "1" to the layout name when it is imported if a duplicate layout name already exists.

Export.   The **Export** button is used to export the layout and save it as a .PLF file.

ZOOM

Another feature provided by *P3e* for modifying the on-screen Gantt chart is the **Zoom** feature. This feature allows you to modify the magnification of the information provided on the screen. Zooming in on a Gantt chart (Figure 5–1) displays a shorter range of time, or it makes the information appearing in the bar chart portion of the screen appear larger. To use the zoom-in feature, click on the **Zoom In** button (Figure 5–1) of the **Activity Toolbar.** Zooming out on a Gantt chart (Figure 5–1) displays a larger range of time, or it makes the information appearing in the bar chart portion of the screen appear smaller. To use the zoom-out feature, click on the **Zoom Out** button (Figure 5–1) of the **Activity Toolbar.**

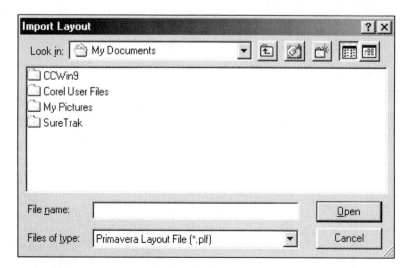

Figure 5–4   Import Layout Dialog Box

COLUMNS

Look at the Columns and Column title location in Figure 5–5. To modify the on-screen columns configuration, select **Columns** from the **View** main pull-down menu. The **Columns** dialog box will appear (Figure 5–6). Before modifying a column, look at the columns as they are presently configured for the Sample Schedule. Close the **Columns** dialog box to return to the on-screen Gantt chart. Click and drag the mouse on the vertical split bar to expose the columns behind the Gantt chart (Figure 5–7). All the columns appearing in Figure 5–7 are in the **Display Columns** field in Figure 5–6. Columns can also appear without the Gantt chart. From the **View** main pull-down menu, select **Show on Top.** Then select the **Activity Table** option. The activity table screen shows only selected tabular information.

The column title as it appears in Figure 5–7 may be edited. To change a column title, select the title to be changed from the **Display Columns** field of the **Columns** dialog box (Figure 5–6). Note that the **Activity ID** title has been selected. Then click on the **Edit Title** button of the

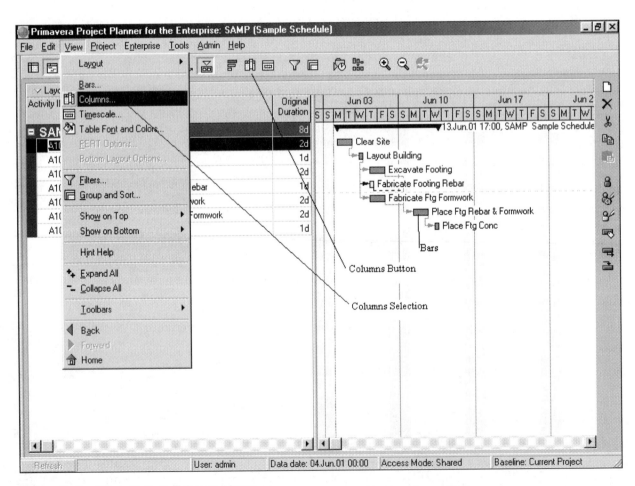

Figure 5–5   **View** Main Pull-Down Menu

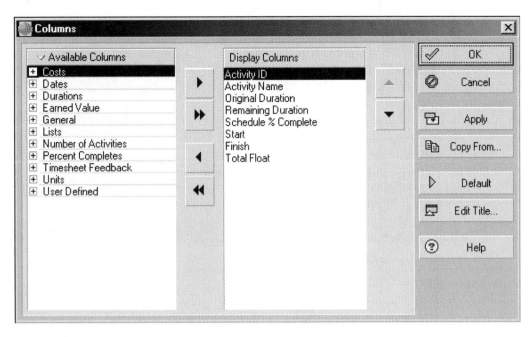

Figure 5–6   Columns Dialog Box

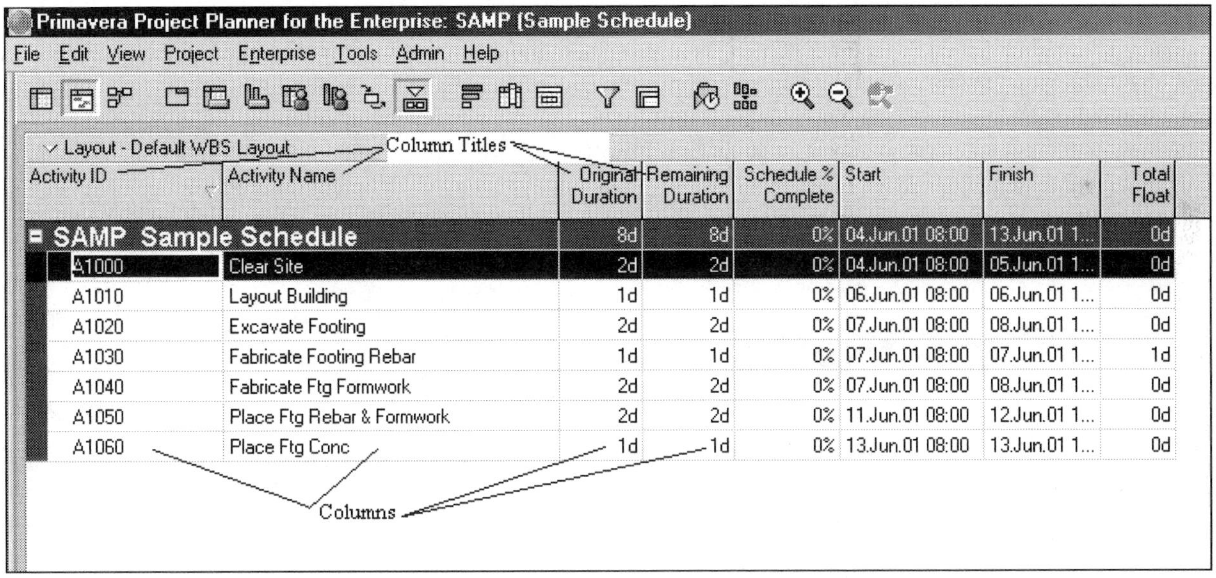

Figure 5–7   Sample Schedule—Columns Exposed

Columns dialog box, and the **Edit Column Title** dialog box will appear (Figure 5–8). Here the **New Title,** column **Width,** and **Alignment** of the selected column can be changed.

To add a new column to the existing display (Figure 5–7), the column must be added to the **Display Columns** field of the **Columns** dialog box (Figure 5–6). Click on the **Available Columns** button and then on the **Expand All** option. Now, all possible *P3e* new column titles will appear

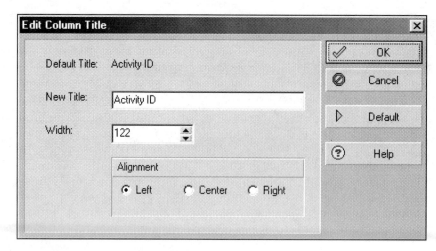

Figure 5–8   Edit Column Title Dialog Box

(Figure 5–9). To add a new column from the **Available Columns** (non-displayed) field to the **Displayed Columns** field, first select the column to be moved **(Actual Duration).** Then from the **Display Column** field, select the column that the moved column will appear after, in this case the **Remaining Duration** title is selected. Then click on the arrow button pointing toward the **Display Column** field. Note that in Figure 5–9, the **Actual Duration** column title now appears after the **Remaining Duration** title. Click the **OK** button of the **Columns** dialog box. Figure 5–10 shows the exposed columns of the Sample Schedule with the new column **(Actual Duration)** appearing after the **Remaining Duration** column.

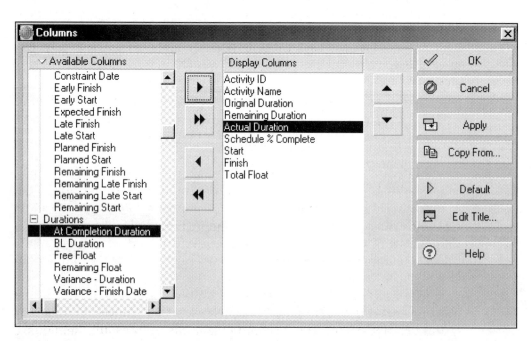

Figure 5–9   Columns Dialog Box—New Column Added

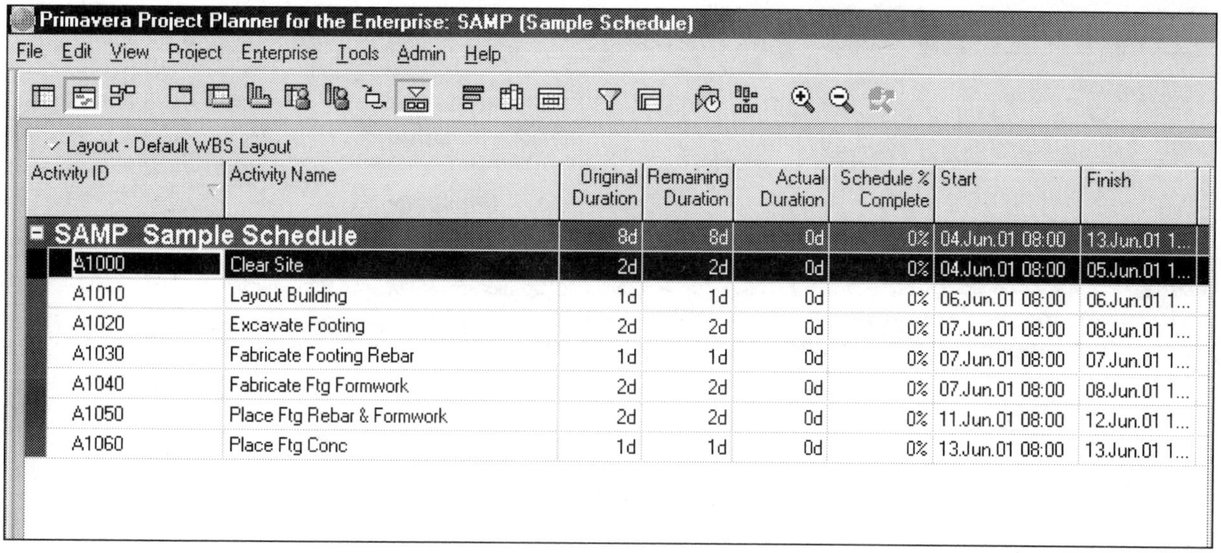

Figure 5–10   Sample Schedule—Column Added

BARS

Look at the bars location in Figure 5–10. To modify the on-screen bars configuration, select **Bars** from the **View** main pull-down menu (see Figure 5–5). The **Bars** dialog box will appear (Figure 5–11).

You can use the **Bars** dialog box to add or delete bars, specify which bars are visible, position of the bars, and modify the bar pattern or colors of the Gantt chart. Note, from Figure 5–11, that there are presently three bars checked as visible for display on the Sample Schedule screen. The Actual Work, Remaining Work, and Critical Remaining Work bars are checked under the **Display** column. These columns will be the visible bars on-screen. By scrolling down the available bar options in Figure 5–11, the Current Bar Labels and the Summary Bar are also checked to be visible. The **Bars** dialog box (Figure 5–11) fields are:

Display.   A mark in the checkbox indicates that the specified bar will be visible in the on-screen Gantt chart. The **Critical Remaining Work** bar (Figure 5–11) is shown in Figure 5–13 since there is a mark in its **Display** checkbox.

Name.   Click the desired bar to select it. Double click the **Name** field to change the bar name.

Timescale.   The **Timescale** field represents the timescale data that the bar represents. The **Timescale** data bar options are: Remaining Bar, Current Bar, % Complete Bar, Plan Bar, Actual Bar, Baseline Bar,

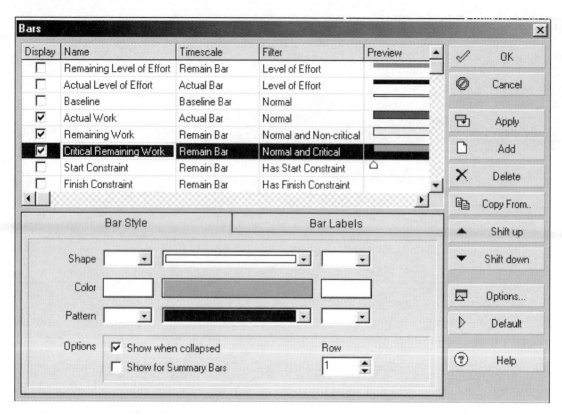

Figure 5–11   Bars Dialog Box

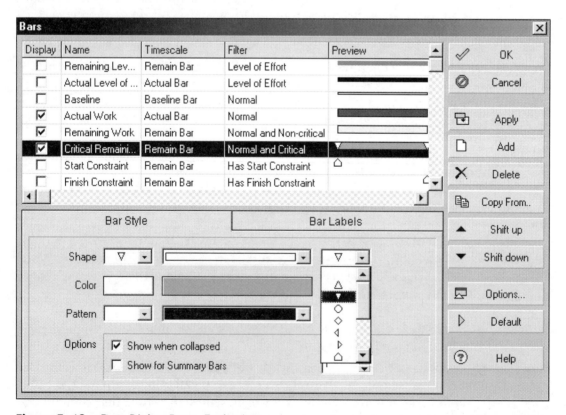

Figure 5–12   Bars Dialog Box—Endpoints

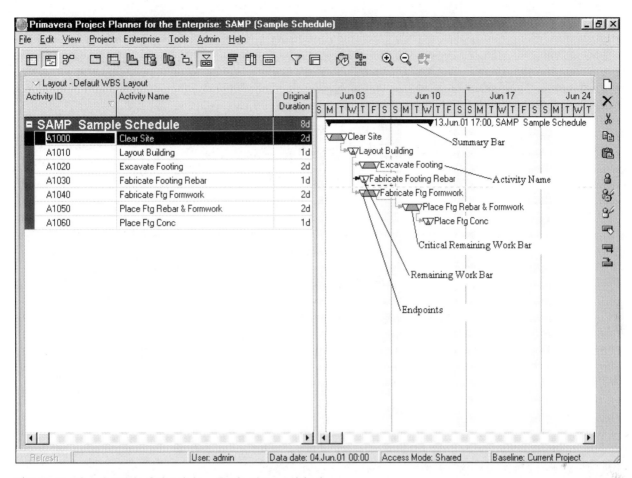

Figure 5–13   Sample Schedule—Endpoints Added

Early Bar, Late Bar, Float Bar, and Negative Float Bar. Double click the **Timescale** to change it.

Filter.   The **Filter** field denotes the possible filter applied to the data item represented by the bar type. The pre-configured *P3e* **Filter** activity data item options are: Completed, Critical, Has Finish Constraint, Has Start Constraint, In Progress, Level of Effort, Milestone, Negative Float, New Feedback to Review, New What-If, Non-critical, Normal, Not Started, and Status to Review. New filters may be added to these pre-configured *P3e* filters.

Preview.   This shows how the bar will appear in the on-screen Gantt chart.

The **Bars** dialog box (Figure 5–11) has two tabs for modifying the display options of a selected bar in the current Gantt chart. The first tab is the **Bar Style** tab. The **Bar Style** tab options can be used to modify the bar and endpoints **Shape, Color,** and **Pattern**. The center three fields are for the bar, and the three fields on both sides of the center fields are for the endpoints. Note that in Figure 5–12, a downward pointing triangle is selected for the endpoints for the **Critical Remaining Work** bar. Figure 5–13 shows the results on the Sample Schedule with the endpoints modified. Compare Figure 5–13, after the endpoint changes, with Figure 5–5 made before the endpoint changes.

The second tab of the **Bars** dialog box is the **Bar Labels** tab appearing in Figure 5–14. The **Bar Labels** tab is used to change the bar labels in the current Gantt chart. Note that, in Figure 5–14, the bar **Current Bar Labels** is selected for evaluating the bar label for the Sample Schedule. The **Bar Labels** tab (Figure 5–14) fields/buttons are:

Position.   Double click this field to change the position of the bar label. The possible positions are: left, right, top, bottom, and center.

Label.   Double click this field to choose a database field for the selected bar's label. Note that in Figure 5–15, the bar **Label** has been changed from Activity Name (Figure 5–13) to Early Start.

Add.   Click the **Add** button (Figure 5–14) to add a bar label to the selected bar. Even though the bar **Current Bar Labels** is selected for modifying the bar label for the Sample Schedule, any of the bars can have a label attached.

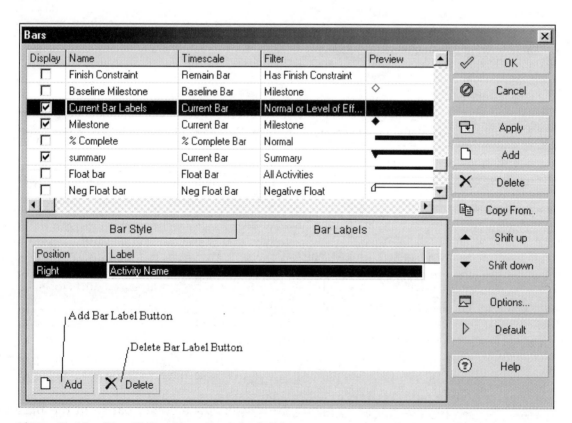

Figure 5–14   Bars Dialog Box—**Bar Labels** Tab

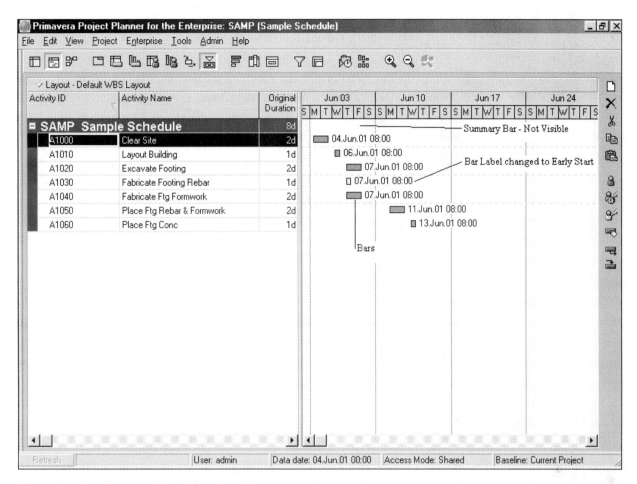

Figure 5–15   Sample Schedule—Bar Changes Made

Delete. Click the **Delete** button (Figure 5–14) to delete a bar label from the selected bar.

The **Bars** dialog box (Figure 5–14) has a number of buttons on the right side of the dialog box for manipulating changes made to the dialog box. The **Bars** dialog box (Figure 5–14) buttons are:

OK. Accepts changes made to the **Bars** dialog box and closes it.

Cancel. Cancels changes made to the **Bars** dialog box and closes it.

Apply. Click this button to preview changes made to the **Bars** dialog box without closing the dialog box.

Add. Click this button to add a bar to the **Bars** dialog box.

Delete. Click this button to delete the selected bar from the **Bars** dialog box. To remove the selected bar from the on-screen display without deleting it, remove the check from the checkbox in the **Display** field for that bar.

Copy From.   To apply the Gantt settings from another layout to the current activity layout, click this button.

Shift Up.   Click this button to move the selected bar up one position in the **Bars** dialog box.

Shift Down.   Click this button to move the selected bar down one position in the **Bars** dialog box.

Options.   This button will be discussed in the upcoming sections on relationship lines and horizontal lines in this chapter.

Default.   Click this button to apply *P3e*'s default Gantt chart settings.

Compare Figure 5–15 to Figure 5–13, note the Summary Bar is no longer visible.

RELATIONSHIP LINES

Another feature in *P3e* to modify the on-screen Gantt chart is **Relationships**. As discussed earlier, the successor and predecessor relationships assigned can be shown on-screen. This is very handy for evaluating the logic relationships. To actually see all the relationship lines on the screen rather than trying to figure them out from some table or dialog box makes the logic relationships much easier to understand. To do this, click on **View** and select **Bars**. Then, from the **Bars** dialog box, select the **Options** button and the **Bar Chart Options** dialog box will appear (Figure 5–16). Click on the **Show Relationships** checkbox to select it. Figure 5–17 is the Sample Schedule with relationships showing.

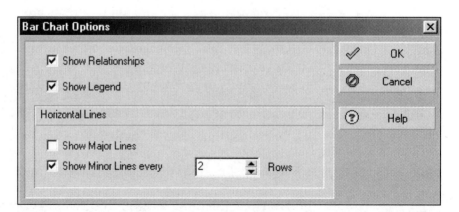

Figure 5–16   Bar Chart Options Dialog Box

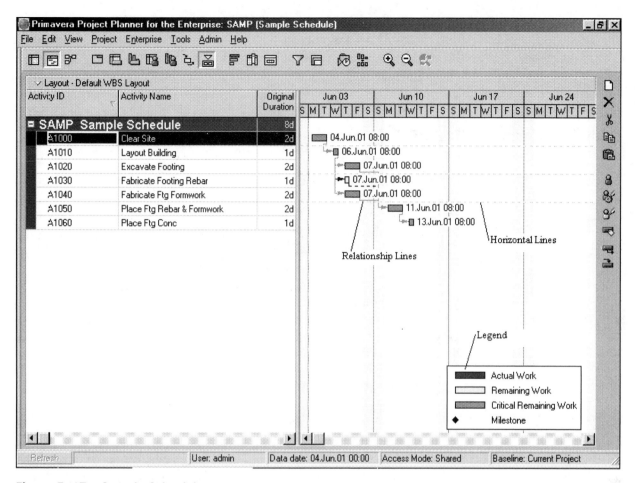

Figure 5–17 Sample Schedule

Compare Figure 5–17 with the relationships showing to Figure 5–15 without relationships.

SIGHT LINES

A *P3e* formatting option that makes the on-screen bar chart and hard copy printouts much easier to read is sight lines. The on-screen horizontal lines are labeled in Figure 5–17. The horizontal lines make it easier to relate the bars to the activity rows of the on-screen Gantt chart. To change the horizontal on-screen lines, select **Bars** from the **View** main pull-down menu. Then from the **Bars** dialog box, select the **Options** button and the **Bar Chart Options** dialog box will appear (Figure 5–16). Under the **Horizontal Lines** portion of the dialog box, the on-screen horizontal lines can be modified. Compare Figure 5–17, with the **Show Minor Lines every** set at 2, to Figure 5–13, with this setting at 5.

Another handy feature of the **Bar Chart Options** dialog box (Figure 5–16) is the **Show Legend** option. Note that when this option is selected the bar legend will appear on-screen (Figure 5–17).

FONT, ROW HEIGHT, AND COLOR

The ability to change the font size for on-screen text is a handy feature for graphic presentation modifications. To change the on-screen text font, select the **Table Font and Colors** option from the **View** main pull-down menu. The **Table Font and Colors** dialog box will appear (Figure 5–18). Select the **Font** button to make modifications.

Changing the row height provides more spacing between activities. Another advantage of being able to change the row height is it enables you to adjust the number of activities on a single page for getting hard-copy prints. Compare Figure 5–17, with the **Row Height** set at 18, to Figure 5–19, with the **Row Height** set at 30.

A *P3e* formatting option that gives the on-screen bar chart and hard-copy printouts a personal touch is the ability to change background colors. The ability to change the color of *P3e* components is a powerful tool in presentations and helps create a company image. To change the background screen color scheme select **Table Font and Colors** from the **View** main pull-down menu. The **Table Font and Colors** dialog box will appear (Figure 5–19).

Click on the **Color** button to reveal a color palette for selecting the on-screen background color. Figure 5–20 is an example of the *P3e* screen changed in color.

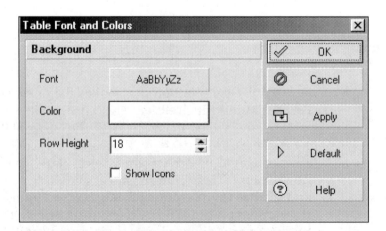

Figure 5–18   Table Font and Colors Dialog Box

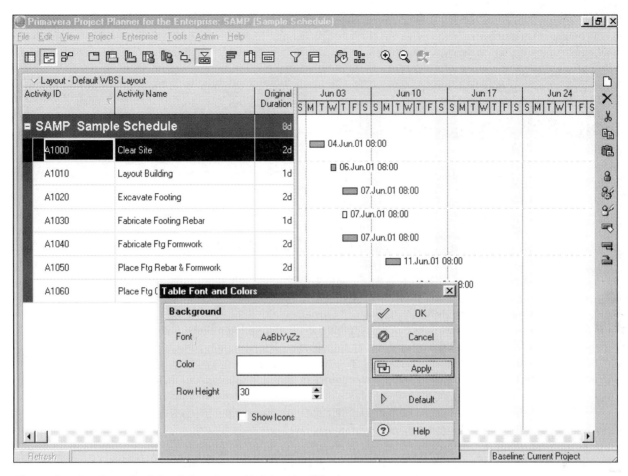

Figure 5–19   Sample Schedule—Row Height Modified

ACTIVITY CODES

Activity codes and their values allow you to filter, group, sort, and report activity information according to your particular needs. *P3e* supports two types of activity codes, global activity codes and project activity codes. You can assign global activity codes and values to activities in all projects. You can assign project activity codes and values to activities only in the project for which the codes were created.

Sometimes it is useful to organize the project by another sort rather than by activity ID (Figure 5–20). For instance, you may want to look at the project by crew—laborer crew versus carpentry crew. Click on the **Enterprise** main pull-down menu; select **Activity Codes** and the **Activity Codes** dialog box will appear (Figure 5–21). In the **Activity Codes** dialog box (Figure 5–21), click on either the **Global** or **Project** checkboxes to

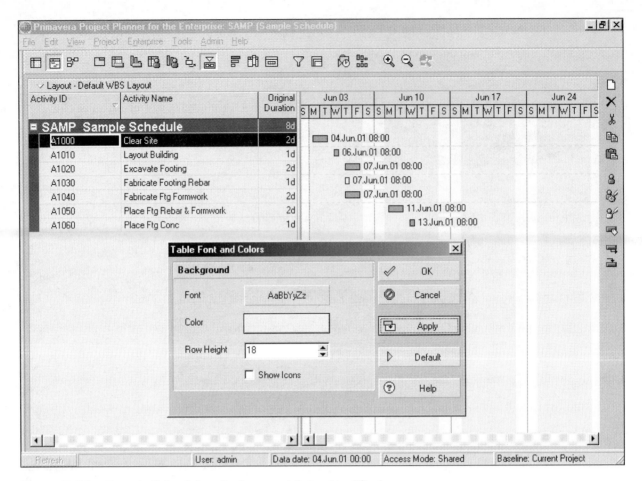

Figure 5–20   Sample Schedule—Background Color Modified

make the new activity code either global (assigned to all projects in the database) or project-specific only.

Note that, from Figure 5–21, under the **Select Activity Code**, no sort option has been selected. By clicking the down arrow box, an activity code used for sorting can be selected. When there are no activity codes defined in the system, or you need to add a new activity code, click on the **Modify** button of the **Activity Codes** dialog box and the **Activity Code Definitions** dialog box will appear (Figure 5–22).

Under the **Activity Code Name** field of the **Activity Code Definitions** dialog box (Figure 5–22), a new activity code sort field named Responsibility is created. By clicking the Responsibility **Activity Code** field to select it and then closing the dialog box, the Responsibility code for the SAMP project appears under the **Select Activity Code** field (Figure 5–23).

Figure 5–21   Activity Codes Dialog Box

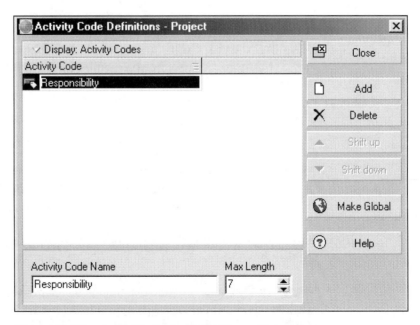

Figure 5–22   Activity Code Definitions Dialog Box

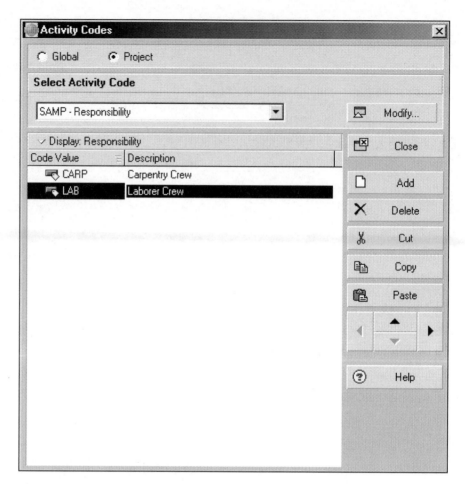

Figure 5–23   Activity Codes Dialog Box—Code Selected

If another *P3e* schedule has been completed and you are pleased with the way the **Activity Codes** are defined, the **Copy** and **Paste** buttons can be used. Many companies use the same or a similar coding system for every project. It is much faster to transfer these codes to each new project than to re-enter them.

Under the **Code Value** field of the **Activity Codes** dialog box (Figure 5–23), select the Add button to create the new activity code sort fields for the Carpenter and Laborer Crews. Under the **Code Value** field type in CARP, and then type Carpentry Crew under the **Description** field. Follow the same procedure to add the new Laborer Crew activity code. Now that the new activity code sort fields have been created, select the **Close** button of the **Activity Codes** dialog box (Figure 5–23) to close the dialog box.

Activity Details.   The next step is to identify each activity by responsibility. Click on the first activity to be identified. In Figure 5–24, Activity A1000 has been highlighted. Click on the **View** main pull-down menu

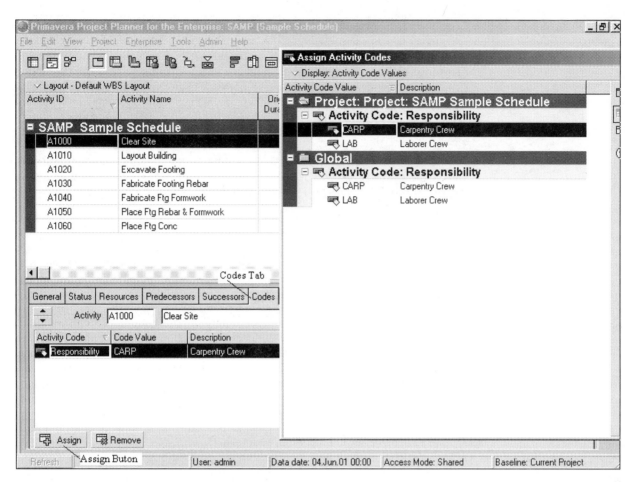

Figure 5–24   Assign Activity Codes Dialog Box

and select **Show on Bottom** and then **Activity Details**. If you access **Activity Details** and not all of the form is exposed, move your cursor to the view control line. The cursor image will change from an arrow to two short double lines. By depressing and holding down the left mouse button, you expose all the **Activity Details** information.

Click the **Codes** tab of the **Activity Details** form to expose the activity code information about the selected activity. Then click on the **Assign** button and the **Assign Activity Codes** dialog box will appear (Figure 5–24). The CARP **Activity Code: Responsibility** is selected for Activity A1000 of the SAMP project. Note that, by selecting it, the CARP Responsibility now appears on **Activity Details** for Activity A1000 (Figure 5–24). Simply click on the next activity and assign its activity code for Responsibility. Continue the process until all activities have been assigned an activity code.

GROUP AND SORT

To view the on-screen Gantt chart by responsibility, click on the **View** main pull-down menu. Select **Group and Sort** and the **Group and Sort** dialog box will appear (Figure 5–25). Click on the level 1 field under the **Group By** column, and a menu will appear for selecting the group and sort possibilities. In Figure 5–25, Responsibility has been selected for the level 1 sort. Note the levels of sort can go nine levels deep. So, if Activity ID were chosen as the level 2 sort, *P3e* would sort all activities by Responsibility and then by Activity ID. Notice that the groups can be customized by **Font & Colors.** Here, the **Apply** button is clicked to pre-view the change. Click on the **OK** button to accept the sort change.

Compare Figure 5–26, sorted by responsibility, to Figure 5–17, sorted by activity ID. Activities A1000, A1010, A1040, and A1050 were defined as Carpentry Crew responsibilities. Activities A1020, 1A030, and A1060 were defined as Laborer Crew responsibilities. The communications advantages are obvious. The sort capabilities of *P3e* make it a great communications tool. Once the information is input, with minor changes, it can be resorted to provide different members of the project team with the scheduling data sorted for their particular needs.

SUMMARIZE

Sometimes it is useful to summarize the project by sorted information. In this way, instead of looking at hundreds of detail activities, project

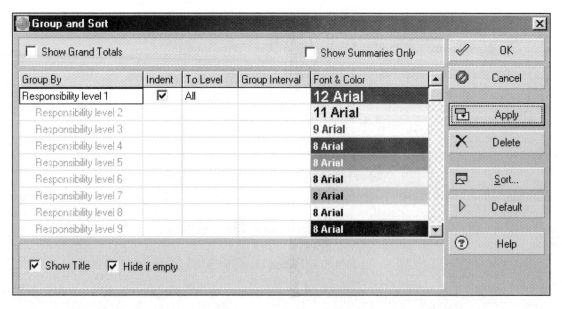

Figure 5–25   Group and Sort Dialog Box

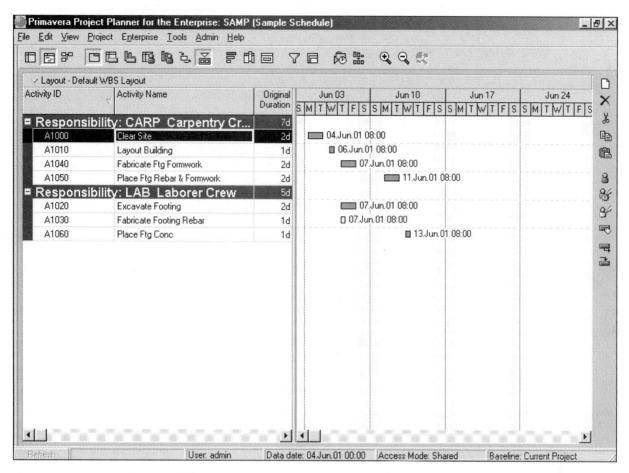

Figure 5-26   Sample Schedule—Grouped by Responsibility

management personnel can look at a few summary activities that are the result of all the detail activities. Note from Figure 5–27 that the Summary bars for Responsibility already appear on the on-screen bar chart. The summary bar can be made visible through the **Bar** options available from the **View** main pull-down menu.

In Figure 5–27 the Sample Schedule in sorted by two groups; Carpentry Crew and Laborer Crew. By clicking the – checkbox beside a group, the group's activities are summarized into a single activity. Compare Figure 5–28, where the group Laborer Crew is summarized, to Figure 5–27, with no summary. *P3e*'s summary feature is a very handy feature for providing only the needed information to different users of the schedule information. To make the hidden activities visible again, click the + checkbox beside the group that was summarized. Another convenient way to control summary bars within *P3e* is to use the **Expand All** and **Collapse All** options available for the **View** main pull-down menu.

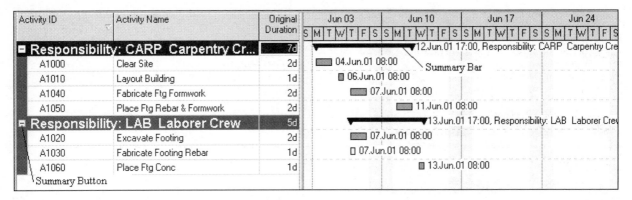

Figure 5–27   Sample Schedule—Summary Bars

FILTER

The **Filters** option can be useful for filtering the project activities and for showing on-screen and hard-copy printouts of only a portion of the activities sorted by some identifier. In this way, instead of looking at hundreds of detailed activities, project management personnel can look at only the activities meeting the sort criteria. For example, you may want to look at only Carpentry responsibility of the Sample Schedule. Click on the **View** main pull-down menu and select **Filter** (see Figure 5–1). The **Filters** dialog box will appear (Figure 5–29).

As can be seen from the **Filters** dialog box (Figure 5–29), *P3e* has a number of predefined filter options. Also, this dialog box can be used to define a new filter criterion. To create a new filter, click on the **New** button of the **Filters** dialog box (Figure 5–29) and the **Filter** dialog box will appear (Figure 5–30). Note that the new filter, Responsibility, is defined under **Filter Name**. Three fields must be used to define the filter criteria. They are:

- **Parameter.**   By clicking the down arrow in the open field under the **Parameter** column, a menu window will open for selection of the filter criteria. The Responsibility activity code, as defined earlier in this chapter, will appear as one of the options available from the menu. Click on Responsibility to select it.
- **Is.**   **Is** establishes the relationship between criteria defined in the **Parameter** field and the **Value** field. The options available for the **Is** field depend upon the **Parameter** selected.
- **Value.**   The **Value** field is used to define which option or options of the **Parameter** to make visible on the screen. When the activity code of Responsibility was created earlier in the chapter, two values for Responsibility were created. The values were Carpentry Crew (CARP) and Laborer Crew (LAB). Under the **Value** field (Figure 5–30) the CARP value was selected.

Activity ID	Activity Name	Original Duration	Jun 03	Jun 10	Jun 17	Jun 24
			S M T W T F S	S M T W T F S	S M T W T F S	S M T W T F S
⊟ Responsibility: CARP Carpentry Cr...		7d		▼12.Jun.01 17:00, Responsibility: CARP Carpentry Cre		
A1000	Clear Site	2d	▭ 04.Jun.01 08:00			
A1010	Layout Building	1d	▪ 06.Jun.01 08:00			
A1040	Fabricate Ftg Formwork	2d	▭ 07.Jun.01 08:00			
A1050	Place Ftg Rebar & Formwork	2d	▭ 11.Jun.01 08:00			
⊞ Responsibility: LAB Laborer Crew		5d	▼13.Jun.01 17:00, Responsibility: LAB Laborer Cre			
			Summarized Group			

Figure 5–28   Sample Schedule—Summarized

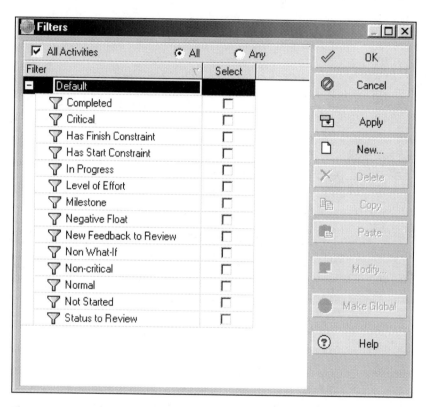

Figure 5–29   **Filters** Dialog Box

The new filter, Responsibility, has been created that will limit activities appearing on the screen to only the activities where responsibility is limited to the carpentry crew. Click the **OK** button of the **Filter** dialog box (Figure 5–30) to save the new filter. Note that when the new filter is saved, you will be returned to the **Filters** dialog box (Figure 5–31). Now the new user-defined filter Responsibility appears at the bottom of the possible filter options. By clicking in the **Select** checkbox by Responsibility, and then clicking the **OK** button, you are returned to the Gantt chart with the new filter criteria in effect. Compare Figure 5–32, which is filtered to limit visible activities only to Carpentry Crew activities, to Figure 5–27, which has no filter limit upon visible activities.

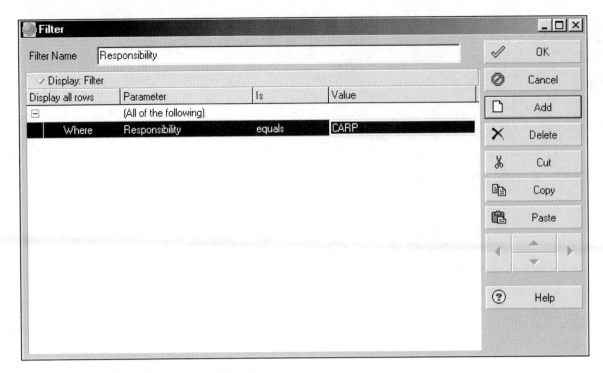

Figure 5–30   Filter Dialog Box—Filter Added

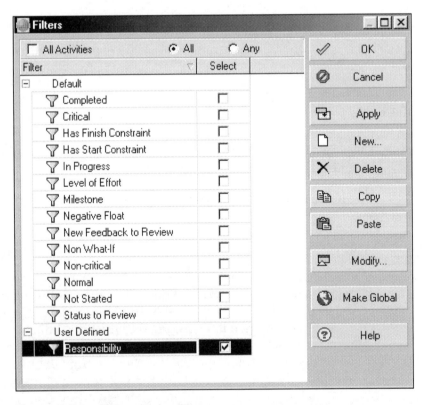

Figure 5–31   Filters Dialog Box

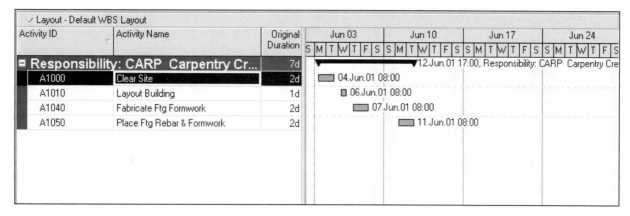

Figure 5–32   Sample Schedule—Filtered

EXERCISES

1. Modify the Zoom feature of the on-screen *P3e* Gantt chart created in Exercise 1 of Chapter 4.
2. Modify the Columns feature of the on-screen *P3e* Gantt chart created in Exercise 2 of Chapter 4.
3. Modify the Bars feature of the on-screen *P3e* Gantt chart created in Exercise 3 of Chapter 4.
4. Modify the Sight Lines feature of the on-screen *P3e* Gantt chart created in Exercise 4 of Chapter 4.
5. Create an Activity Codes feature for the on-screen *P3e* Gantt chart created in Exercise 5 of Chapter 4.
6. Use the Organize feature for the on-screen *P3e* Gantt chart created in Exercise 6 of Chapter 4.
7. Use the Summarize feature for the on-screen *P3e* Gantt chart created in Exercise 7 of Chapter 4.
8. **Small Commercial Concrete Block Building—On-Screen Gantt Chart**
   Using the on-screen Gantt chart for the small commercial concrete block building created in Exercise 9 of Chapter 4, use the following *P3e* features to modify its appearance:
   A. Zoom
   B. Columns
   C. Bars
   D. Sight lines
   E. Row height

9. **Large Commercial Building—On-Screen Gantt Chart**
   Using the on-screen Gantt chart for the large commercial building created in Exercise 9 of Chapter 4, use the following *P3e* features to modify its appearance:
   A. Screen colors
   B. Activity codes
   C. Organize
   D. Summarize
   E. Filter

6

Resources

Objectives

Upon completion of this chapter, you should be able to:

- Modify the resource dictionary
- Assign activity resources
- Use a resource profile
- Use a resource spreadsheet
- Define driving resources
- Set resource limits
- Level resources

NECESSITY OF CONTROLLING RESOURCES

The contractor must be able to control resources—labor hours, bulk materials, construction equipment, and permanent equipment. The ability to get the greatest bang for the buck in putting the resources in place usually determines a contractor's success. Control of resources involves:

- Optimizing resources
- Paying attention to cost and time
- Controlling waste
- Paying attention to detail
- Preplanning
- Paying attention to efficiencies

Resource loading the schedule (defining project requirements for people, materials, and equipment) is an effective way to control resources. Loading the labor resources identifies weeks in advance the exact activities to be worked on as of a particular day as well as the crafts and number of workers per craft that are required. Knowing the subcontractors' labor plans is also an effective way to measure subcontractor performance.

The concepts of resource limits and leveling need to be introduced at this time. A *limit* is the maximum amount of a resource available at one time.

If a person serves a unique function on a project, the limit is one. If five workers of one type are available at any one time, then the limit is five. Obviously, resource limitations impact the scheduling of the project and the relationship of activities to each other.

Resource *leveling* is the redistribution of resources to eliminate resource conflict. If a resource is overallocated (more is assigned than is available for a given time period), *P3e* can be directed to reschedule the activities (modify the activity logic or durations) so that the resources are not over-committed (the resource does not exceed its limit).

Remember that time is money. A contractor who uses only a cost system to control resources is not getting the complete picture in the effective use of the resources. The entire scheduling process should be looked at from the point of view of efficient use of time, money, and resources.

Define Resources and Requirements

The scheduler has some flexibility in determining the resource requirements of a project. Resources are anything needed for the project execution that are paid for to acquire. Resources are typically thought of as labor, material, or construction equipment. In this chapter, only direct labor will be analyzed. We begin by defining the labor requirements of each activity; then we evaluate the requirements of the entire project and refine it if necessary.

Defining Resources

The resources must be defined before the requirements of each activity can be addressed. Click on the **Enterprise** main pull-down menu; then click on the **Resource Codes** option (Figure 6–1). The **Resource Codes** dialog box will appear (Figure 6–2). *P3e* enables you to categorize resources using resource codes. With potentially many resources being used across many projects, *P3e*'s resource codes provides a way for sorting resources that you need for grouping and summarization purposes. To create a new resource code, select the **Modify** button of the **Resource Codes** dialog box (Figure 6–2) and the **Resource Code Definitions** dialog box will appear (Figure 6–3). Here, the **Add** button was clicked, and the resource code, **Labor Classes – Own Forces** was created. To create another resource code, simple click the **Add** button again and create the new code.

Now, with the resource codes defined, the next step is to define the resources themselves. Click on the **Enterprise** main pull-down menu, and select **Resources**. The **Resources** screen will appear (Figure 6–4). To

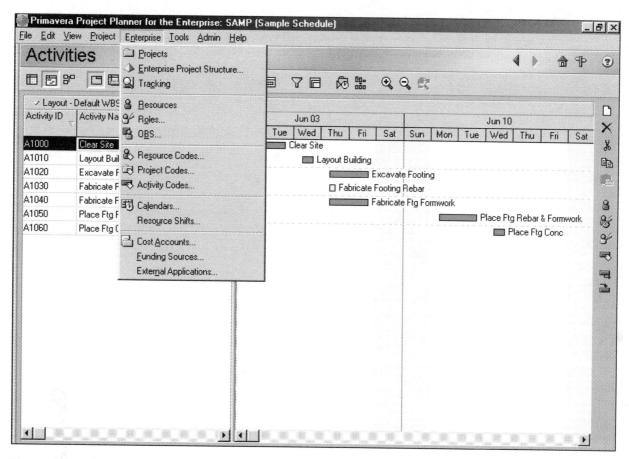

Figure 6–1   Sample Schedule

add a new resource click on the add button and the **New Resource Wizard** will be initiated (Figure 6–5). In the first window of the **New Resource Wizard** is the **Resource ID and Name** (Figure 6–5). The resource with a **Resource Name** of **Carpenter Class 1** and a **Resource ID** of **Carp 1** is input. Select the **Next** button to input more information about the newly created resource.

In the second window of the **New Resource Wizard** is the **Labor Classification** window (Figure 6–6). The new resource (**Carpenter Class 1**) is defined as a **Labor** resource as opposed to a **Nonlabor** resource. This distinction is critical for internal *P3e* calculations used in progressing activities. Select the **Next** button to input more information about the newly created resource.

From the last window of the **New Resource Wizard** select the **Finished** button and the creation of the new resource (**Carpenter Class 1**) is complete. Figure 6–7 shows all the new resources that will be used for the Sample Schedule.

Figure 6–2   Resource Codes Dialog Box

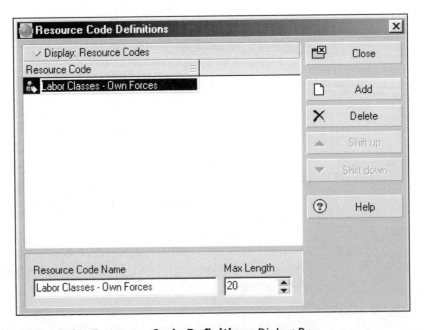

Figure 6–3   Resource Code Definitions Dialog Box

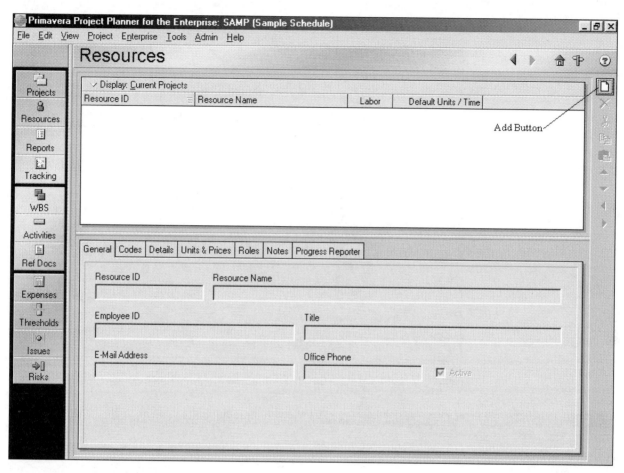

Figure 6–4   Resources Screen

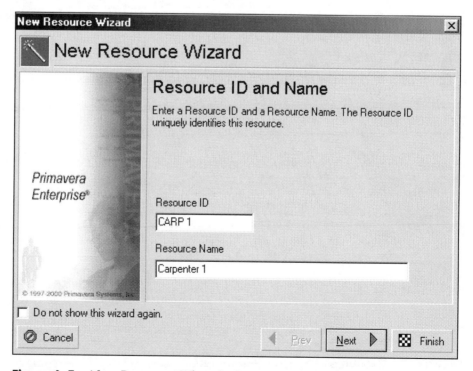

Figure 6–5   New Resource Wizard—Resource ID and Name

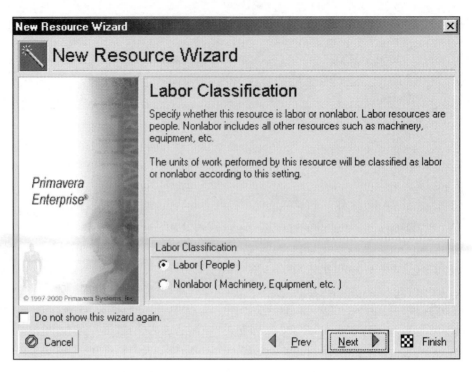

Figure 6–6 New Resource Wizard—Labor Classification

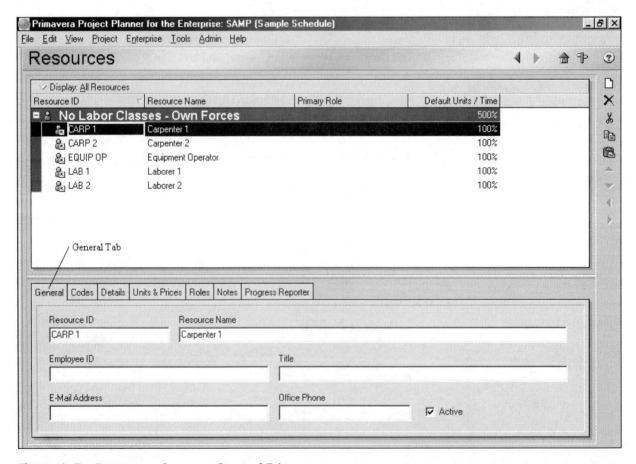

Figure 6–7 Resources Screen—General Tab

Note that at the bottom of the **Resources** screen (Figure 6–7) there are numerous tabs containing the information that was defined by using the **New Resource Wizard**. Figure 6–7 shows the **General** tab of the **Resources** screen.

Figure 6–8 shows the **Details** tab of the **Resources** screen. Note that in Figure 6–8, the **Auto Compute Actuals** checkbox is unchecked. If this checkbox is checked, *P3e* recalculates durations based upon the resources and their budgets input per activity. If you want to input durations and not have *P3e* recalculate and change them based upon resource production, it is important that this checkbox not be checked. For the purposes of this book, the **Auto Compute Actuals** function of *P3e* will not be used.

ASSIGN RESOURCES

When the crafts (resources) have been defined to *P3e*, the requirements of each activity can be specified. To input resource data, you need to

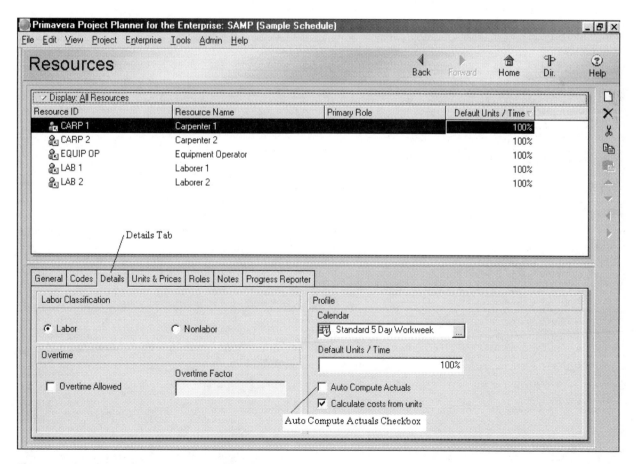

Figure 6–8   Resources Screen—**Details** Tab

access the **Resources** tab of the **Activity Details** window. To reach this window, click on the **Activity Details** button. Then, select the **Resources** tab (Figure 6–9). See Table 6–1 for a listing of the resources and the units per day to be input into the sample schedule.

Resources Dialog Box

Use the **Resources** tab of the **Activity Details** window (Figure 6–9) to specify activity resource requirements within *P3e*. To add a resource requirement to an activity, click on the **Add Resource** button (Figure 6–9), and the resource listing as previously defined will appear (Figure 6–10). Click on the desired resource and click the **Assign** button. In Figure 6–10, **CARP 1** (**Carpenter 1**) is selected.

To assign more than one unit of the same resource, with the desired resource highlighted, simply click on the **Assign** button again. Note that, in Figure 6–11, there have been two **CARP 1** resources selected.

As you can see from Figure 6–12, you can specify more than one resource per activity. This figure shows the requirements for Activity A1010, Building Layout. **CARP 1**, **CARP 2**, and **LAB 1** are specified.

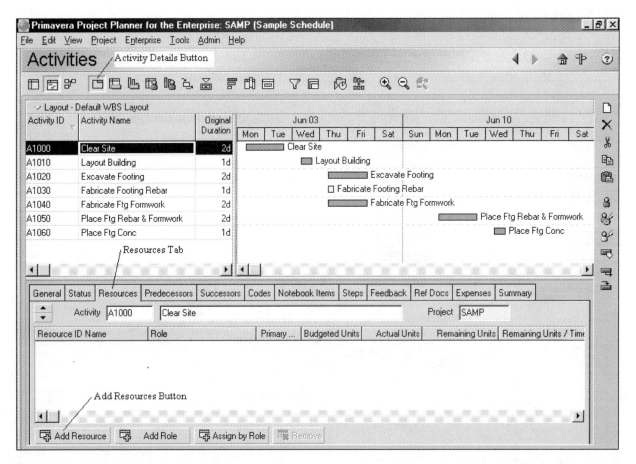

Figure 6–9   Activity Details—Resources Tab

Activity ID	Description	Resource	Units Per Day
1000	Clear Site	CARP 1	1
1010	Building Layout	CARP 1	1
		CARP 2	1
		LAB 1	1
1020	Excavate Footings	CARP 1	1
		LAB 2	1
		EQUIP OP	1
1030	Fabricate Ftg Rebar	LAB 1	1
		LAB 2	1
1040	Fabricate Ftg Formwork	CARP 1	1
		CARP 2	1
		LAB 2	1
1050	Place Ftg Rebar & Formwork	CARP 1	1
		LAB 2	1
		EQUIP OP	1
1060	Place Ftg Conc	CARP 1	1
		LAB 1	1
		EQUIP OP	1

Table 6–1   Sample Schedule Resource List

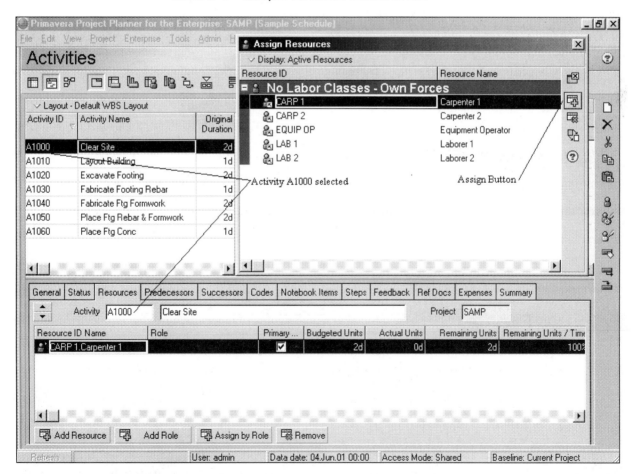

Figure 6–10   Assign Resources Window

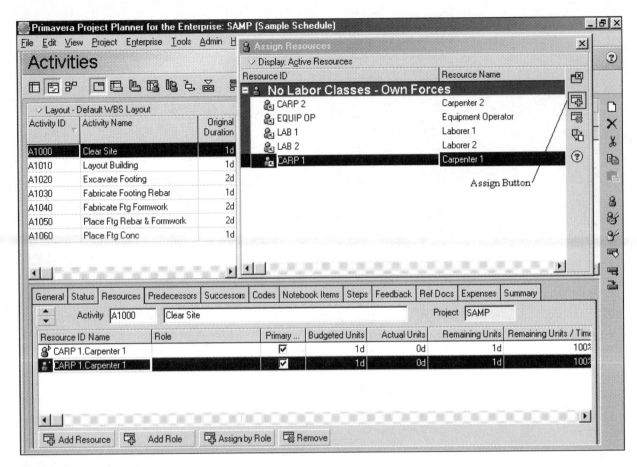

Figure 6–11 Assign Multiples of the Same Resource

Next Activity

Instead of closing the **Assign Resources** window after inputting the resource requirements for each activity, simply click on the next activity and a blank **Assign Resources** window will appear for that activity. This window does not have to be closed until all resource requirements are defined.

RESOURCE PROFILES

The purpose of resource profiles is to create graphical representations of the resource requirements of the project. By placing the resource profile under the bar chart, it makes visual interpretation and analysis simpler.

In the previous sections, the resources were assigned and the individual activity requirements were defined in the **Resources** tab of the **Activity Details** window. Now that all the information is in the system, you can

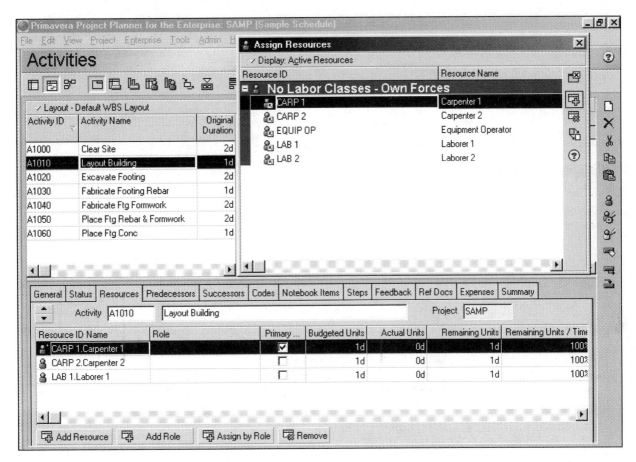

Figure 6–12   Assign Multiple Resources

use the information. To do this, look at resource profiles. To access a total resource profile, click on the **Activity Usage Profile** button of the Activity Toolbar (Figure 6–13). The **Activity Usage Profile** will appear at the bottom of the screen.

A profile is a side view or cross section. A *resource profile* is a graphical representation of resources across time. The **Activity Usage Profile** is a graphical representation of <u>all</u> activity resources. Note that a handy option is provided on the **Activity Usage Profile** for viewing either **All activities** or only **Selected activities.** All resources (all carpenters, laborers, and equipment operators, etc.) as defined by the resource dictionary and on the **Resources** dialog box of the individual activities are shown.

The profile shows:	Mon and Tue	2 days	1 craftsperson
	Wed	1 day	3 craftspersons
	Thu	1 day	8 craftspersons
	Fri	1 day	6 craftspersons
	Sat and Sun	2 days	0 craftspersons (weekend)
	Mon to Wed	3 days	3 craftspersons

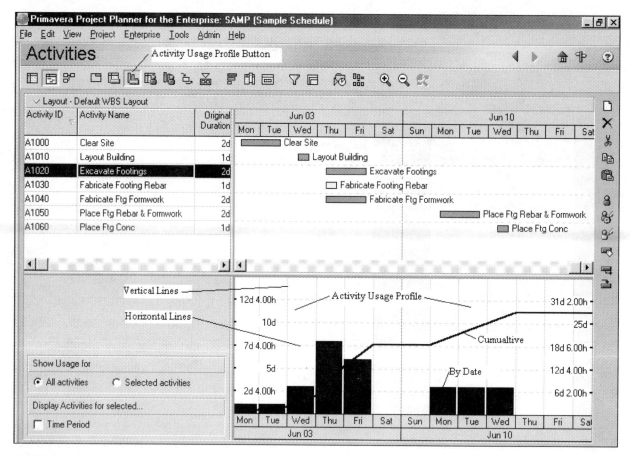

Figure 6–13   Activity Usage Profile

Note again that the **Activity Usage Profile** is a graphical representation of <u>all</u> activity resources. To access an individual resource profile, click on the **Resource Usage Profile** button of the Activity Toolbar (Figure 6–14). The **Resource Usage Profile** will appear at the bottom of the screen. Note that , in Figure 6–14, the profile shows only the requirements for **CARP 1.**

The profile shows:	Mon to Wed	3 days	1 CARP 1
	Thu and Fri	2 days	2 CARP 1
	Sat and Sun	2 days	0 CARP 1 (Weekend)
	Mon to Wed	3 days	1 CARP 1

Obviously, this resource profile is not flat. The wide range of needs from one to eight people in a given week may create a staffing problem. Floats can be used to shift activities around to "flatten" the peak resource requirements.

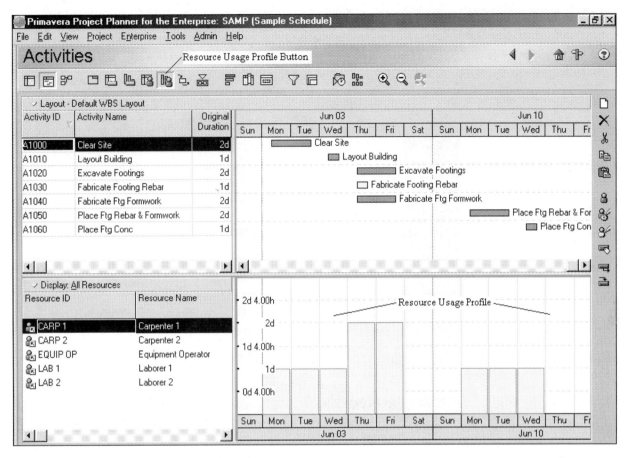

Figure 6–14   Resource Usage Profile

Modify Appearance

The appearance of the **Activity Usage Profile** and the **Resource Usage Profile** can be modified. The two profile types may be modified by using the **Activity Usage Profile Options** or **Resource Usage Profile Options** dialog boxes. Since they are very similar, only the **Activity Usage Profile Options** dialog box will be shown. To access the **Activity Usage Profile Options** dialog box, place your mouse anyplace within the screen of the **Activity Usage Profile** (Figure 6–13) and click the right mouse button. Then select **Activity Usage Profile Options** and the **Activity Usage Profile Options** dialog box will appear (Figure 6–15). This dialog box has two tabs, the **Data** and the **Graph** tabs. The **Data** tab fields (Figure 6–15) are:

> **Display.**   The **Units** option was selected because this chapter deals with resource quantities (units) and not costs. The resource totals (bars) shown in Figure 6–13 are for total craftspersons per day.

> **Filter.**   Only **Labor** resources have been selected.

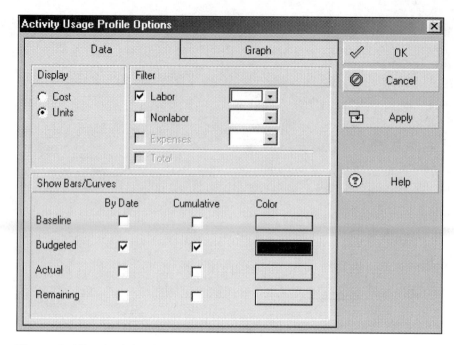

Figure 6–15   Activity Usage Profile Options—Data Tab

Show Bars/Curves. Here four different profiles can be accessed. The options are the **Baseline, Budgeted, Actual,** and **Remaining** resource requirements. In Figure 6–15, only the **Budgeted** resource requirements were selected for viewing in Figure 6–13. By placing the check in the **By Date** checkbox the individual total resource bars are shown by date. Note that, in Figure 6–13, a cumulate curve of total resource requirements is also shown. This was obtained by putting a check in the **Cumulative** checkbox by the **Budgeted** resource requirements

The options that have been selected from the **Graph** tab of the **Activity Usage Profile Options** dialog box (Figure 6–16) are:

Vertical Lines.   Note that, in Figure 6–16, the checkboxes for both **Major** and **Minor** are checked. The results of this are shown in Figure 6–13. The major vertical lines separate weeks. The minor vertical lines separate days.

Horizontal Lines.   The checkbox options of **None** (for not showing horizontal lines), **Dotted,** and **Solid** for horizontal lines are provided. The **Line Color** palate is also provided for changing the horizontal line color. Note that, in Figure 6–16, the dotted horizontal line option is selected. The results of this are shown in Figure 6–13.

Additional Display Options.   Placing a check in the **Show Legend** checkbox will produce an on-screen legend for the **Activity Usage Profile** (Figure 6–13). The **Background Color** palate is also provided for changing the background color of the **Activity Usage Profile**.

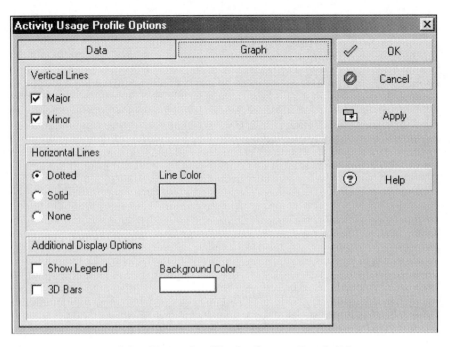

Figure 6–16   Activity Usage Profile Options—Graph Tab

RESOURCE SPREADSHEETS

To access a total resource by activity spreadsheet, click on the **Activity Usage Spreadsheet** button of the Activity Toolbar (Figure 6–17). The **Activity Usage Spreadsheet** will appear at the bottom of the screen. The **Activity Usage Spreadsheet** is a tabular representation of all resources by activity. For example, Activity A1020 is budgeted a total of 3d (days) of labor resources for Thursday and Friday (Figure 6–17). As can be seen from Table 6–1, the resource requirements for Activity A1020 are one craftsperson each from the resources of **CARP 1, LAB 2,** and **EQUIP OP** to equal the 3-day total for Thursday and Friday.

To access an individual resource spreadsheet, click on the **Resource Usage Spreadsheet** button of the Activity Toolbar (Figure 6–18). The **Resource Usage Spreadsheet** will appear at the bottom of the screen. With an individual resource selected (Figure 6–18 shows the requirements for **CARP 1**), only the requirements for that resource are shown in the **Resource Usage Spreadsheet.**

Note on the **Resource Usage Spreadsheet** that a totals line is provided for the resource giving the individual resource total by day. Under the totals line is a listing of all the activities where the resource has been specified. The **CARP 1** resource was specified for Activities A1000, A1010, A1020, A1040, A1050, and A1060. The right side of the **Resource Usage Spreadsheet** shows on which day the resource is used. The total resource requirement for **CARP 1** for Thursday is a 2d total. The 2d total is made up of 1d from Activity A1020 and 1d from Activity A1040.

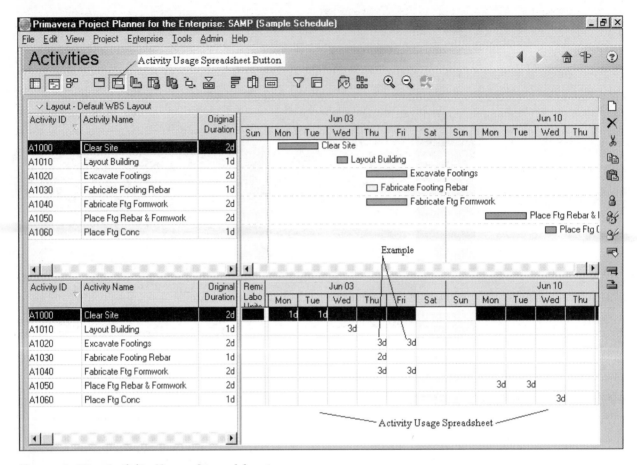

Figure 6–17   Activity Usage Spreadsheet

A very useful feature of the **Resource Usage Spreadsheet** is the **Display Activities for Selected** feature (Figure 6–19). Note that, with **CARP 2** selected and with the **Resource** checkbox also selected, only those activities using the selected resource will appear in the Gantt chart portion of the screen. Note also that, from Table 6–1, **CARP 2** was only specified for two activities (A1010 and A1040). Filtering out the other activities makes the analysis of individual resources easier to visualize.

SET RESOURCE LIMITS

The bottom portion of the **Resources** screen is used to set limits on the availability of a resource. Click on the **Enterprise** main pull-down menu, and select **Resources.** Then on the lower portion of the screen, select the **Units & Price** tab (Figure 6–20). The **Max Units/Time** field is used to set the resource limit. Set the number of units available during each work period (hour, day, week, or month). You can enter a percentage (Figure 6–7) or a numeric value followed by a forward slash (/) and the appropriate time duration. To change from the percentage (Figure 6–7) to the

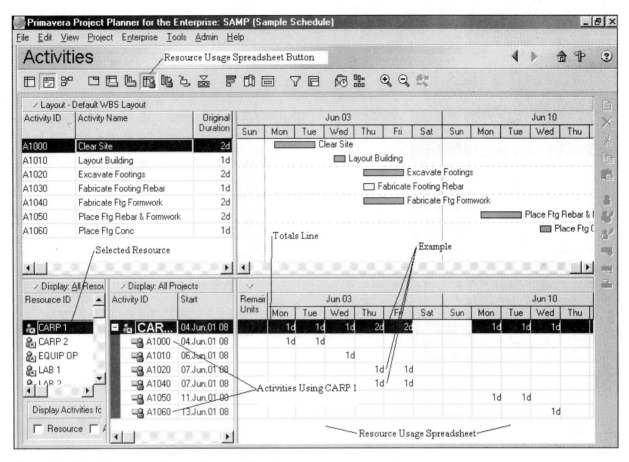

Figure 6–18   Resource Usage Spreadsheet

numeric value, *P3e* must be reconfigured. Select **User Preferences** from the **Edit** main pull-down menu. Select the **Time Units** tab from the **User Preferences** dialog box (Figure 4–36). Under **Resource Units/Time**, click the **Show as units/duration (4h/d)** checkbox. Note that, in Figure 6–20, the **Default Units/Time** is changed to numeric. The maximum units/time (resource limit) of the resource **LAB 2** is set at 2.

To observe the resource limit just placed on **LAB 2,** select the **Resource Usage Profile** button from the **Gantt Chart** screen (Figure 6–21). Note that the horizontal line showing the resource limit (maximum units/time) of the resource **LAB 2** at 2 is visible. To turn this feature on within *P3e*, click the right mouse button within the space of the **Resource Usage Profile** and the **Resource Usage Profile Options** dialog box will appear (Figure 6–22). Note that the **Show Limit** and the **Show Overallocation** checkboxes are checked. The check in the **Show Limit** checkbox causes the limits horizontal line in Figure 6–22 to appear. The check in the **Overallocation** checkbox causes the dark box on Thursday in Figure 6–22 to appear showing that the limit of 2 was exceeded. On Thursday a resource requirement of 3 is shown with the overallocation beyond 2 shown in a different color.

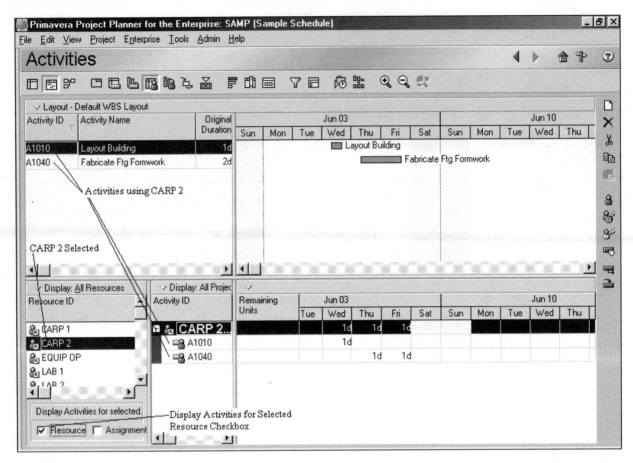

Figure 6–19   Display Activities for Selected Resource Checkbox

LEVEL RESOURCES

In Figure 6–21, the normal limit for **LAB 2** is exceeded. Rearranging priorities of the project so that the resource limit will not be exceeded is called *resource leveling*. Resource leveling means trying to take the peaks out of the profile to lower the overall need for a resource, thus lessening the problems associated with mobilizing and demobilizing personnel. Of course, the first option in trying to level resources is to use positive floats to rearrange activities without prolonging project duration.

Leveling Resources

To begin *P3e*'s leveling process, click on the **Level Resources** button on the Activity Toolbar (Figure 6–21), and the **Level Resources** dialog box will appear (Figure 6–23).

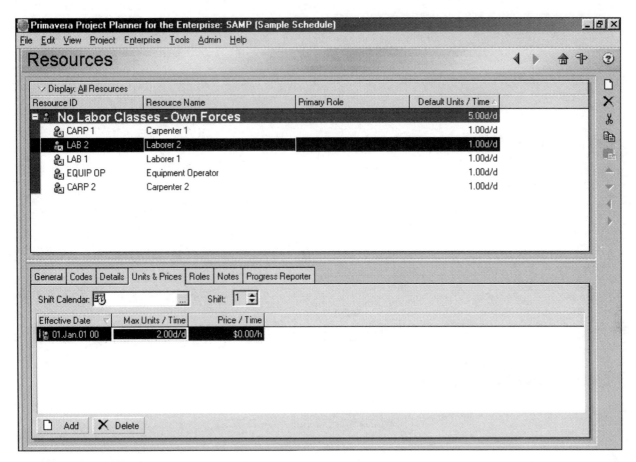

Figure 6–20   Resources Screen—**Units & Prices** Tab

Dialog Box for Resource Leveling

In the **Resource Leveling** dialog box, you have the following options:

Automatically level resources when scheduling.   Click on this box if you want *P3e* to level resources every time it calculates the schedule.

Preserve scheduled early and late dates.   Click on this box if you want to level resources but do *not* want to delay the schedule. If you leave it unchecked, *P3e* will level resources and possibly extend the current calculated project finish date.

Leveling Priorities.   This field identifies priorities in the leveling process. The default is activity priority, but you may change these priorities. Under the **Sort Order** column, if **Ascending** is selected, the most critical activities are leveled first. If **Descending** is selected, the least critical activities get resources first.

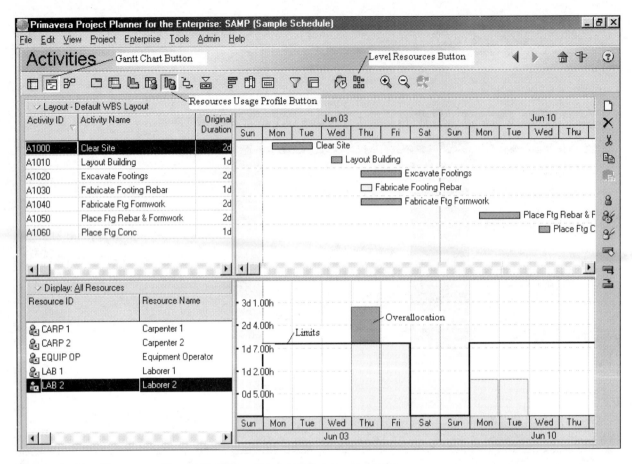

Figure 6–21   Resource Usage Profile—Showing Resource Limits

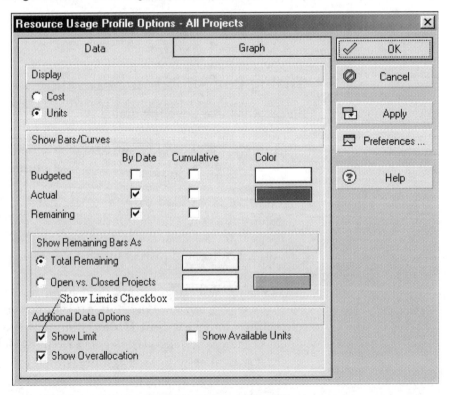

Figure 6–22   Resource Usage Profile Options Dialog Box

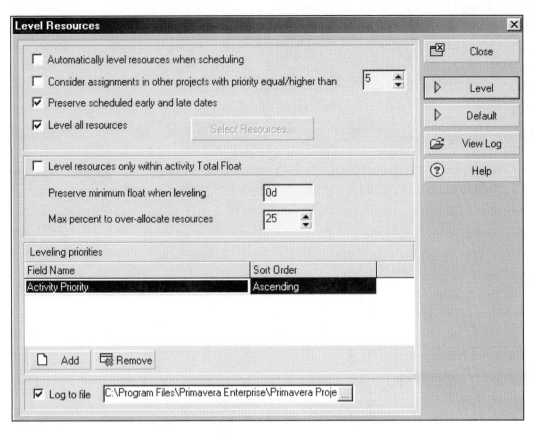

Figure 6–23   Level Resources Dialog Box

You assign priority to activities to tell *P3e* how to resolve resource conflicts when you level resources. For example, suppose you use Total Float and Remaining Duration as the priority-defining items. Activities with the least amount of total float will take priority over those with lots of float. Activities with equal total float will get resources in order of remaining duration (i.e., those scheduled to finish first get resources first).

Level.   Click on the **Level** button to initiate the leveling process by *P3e*.

Results

Figure 6–24 displays the results of the leveling process by *P3e*. Compare Figure 6–24 with Figure 6–21 to see the changes in the resource leveling exercise. Notice that of the activities requiring **LAB 2**, Activity A1040 was the only activity moved and that the project duration was extended by one day. On a larger project with more activities, realtionships, and floats, there is considerable manipulation that can tale place without extending project duration.

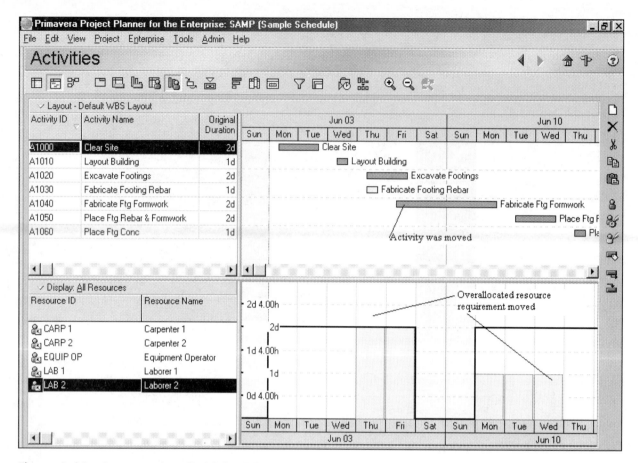

Figure 6-24 Resource Leveled Schedule

EXAMPLE PROBLEM

Table 6–2 is an Activity List with labor resources for a house put together as an example for student use (see the wood-framed house drawings in the Appendix). The on-screen total resource profile (Figure 6–25) was constructed using the tabular list of 28 activities (activity ID, description, and labor classification). Figure 6–26 is a resource profile for the CARPENTR craft. Figure 6–27 is a detailed resource spreadsheet and Figure 6–28 is a total resource spreadsheet.

Activity ID	Activity Description	Labor Classification
A1000	Clear Site	2 – Sub, Site
A1010	Building Layout	2-CARPENTR, 1-CRPN FOR, 1-LAB CL 1, 1-LAB CL2
A1020	Form/Pour Footings	2-CARPENTR, 1-CRPN FOR, 1-LAB CL 1, 1-LAB CL2
A1030	Pier Masonry	2-MASON, 1-LAB CL1, 2-LAB CL
A1040	Wood Floor System	2-CARPENTR, 1-CRPN FOR, 2-CRPN HLP
A1050	Rough Framing Walls	2-CARPENTR, 1-CRPN FOR, 2-CRPN HLP
A1060	Rough Framing Roof	2-CARPENTR, 1-CRPN FOR, 2-CRPN HLP
A1070	Doors and Windows	2-CARPENTR, 1-CRPN FOR, 2-CRPN HLP
A1080	Ext Wall Board	2-CARPENTR, 1-CRPN FOR, 2-CRPN HLP
A1090	Ext Wall Insulation	2 – Sub, Insul
A1100	Rough Plumbing	3 – Sub, Plbg
A1110	Rough HVAC	2 – Sub, HVAC
A1120	Rough Elect	3 – Sub, Elec
A1130	Shingles	4 – Sub, Roof
A1140	Ext Siding	2-CARPENTR, 1-CRPN FOR, 2-CRPN HLP
A1150	Ext Finish Carpentry	2-CARPENTR, 1-CRPN FOR, 2-CRPN HLP
A1160	Hang Drywall	3 – Sub, Drywall
A1170	Finish Drywall	2 – Sub, Drywall
A1180	Cabinets	2 – Sub, Cab
A1190	Ext Paint	3 – Sub, Paint
A1200	Int Finish Carpentry	2-CARPENTR, 1-CRPN FOR, 2-CRPN HLP
A1210	Int Paint	2 – Sub, Paint
A1220	Finish Plumbing	2 – Sub, Plbg
A1230	Finish HVAC	2 – Sub, HVAC
A1240	Finish Elect	2 – Sub, Elec
A1250	Flooring	2 – Sub, Floor
A1260	Grading & Landscaping	2 – Sub, Site
A1270	Punch List	

Table 6–2   Activity List with Labor Resources—Wood Framed House

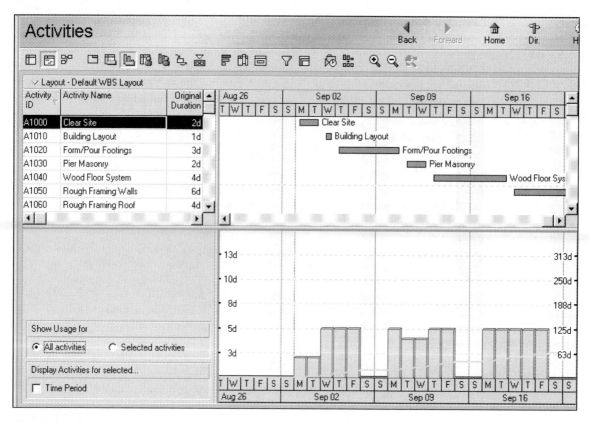

Figure 6–25   All Resources Profile—Wood-Framed House

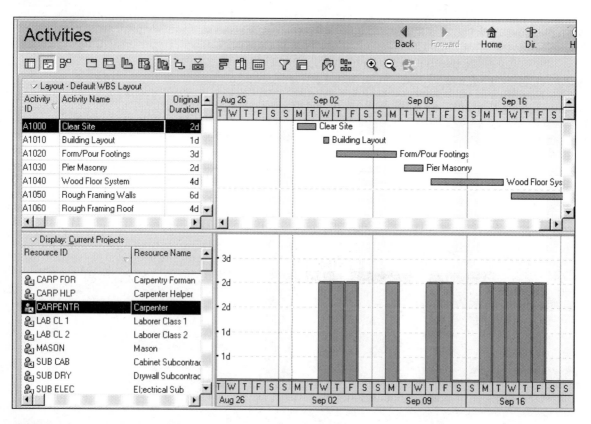

Figure 6–26   Carpenter Profile—Wood-Framed House

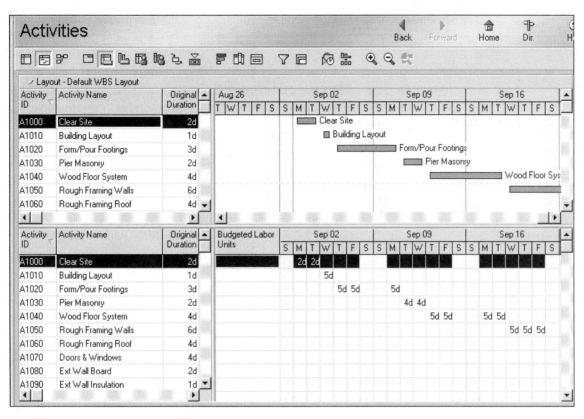

Figure 6–27   All Resources Spreadsheet—Wood-Framed House

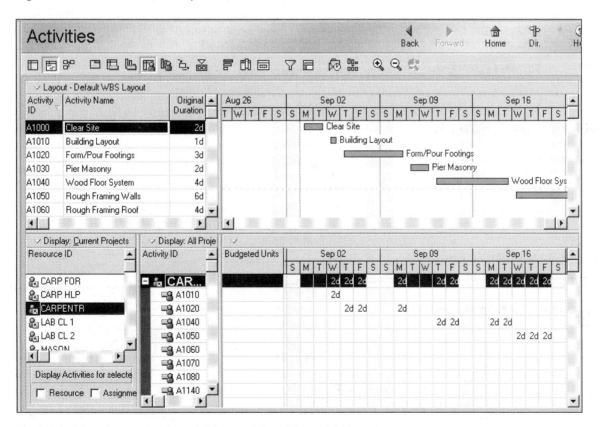

Figure 6–28   Carpenter Spreadsheet—Wood Framed House

EXERCISES

Complete a manual resource profile and resource table for each precedence diagram for Exercises 1 thru 6. Base the resource profile and table upon schedule early start calculations. The labor resources provided in these diagrams are generic.

1.

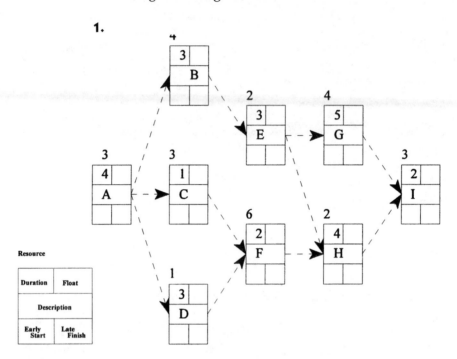

Figure 6–29   Exercise 1—Precedence Diagram with Labor Resource

2.

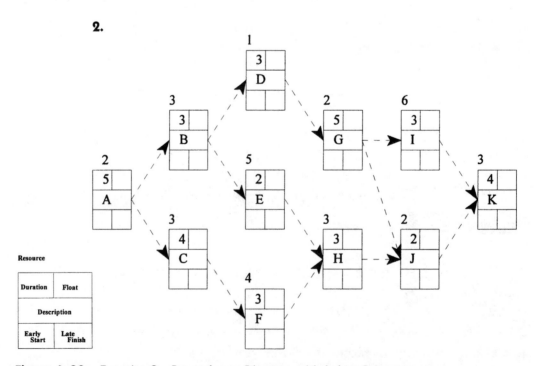

Figure 6–30   Exercise 2—Precedence Diagram with Labor Resource

3.

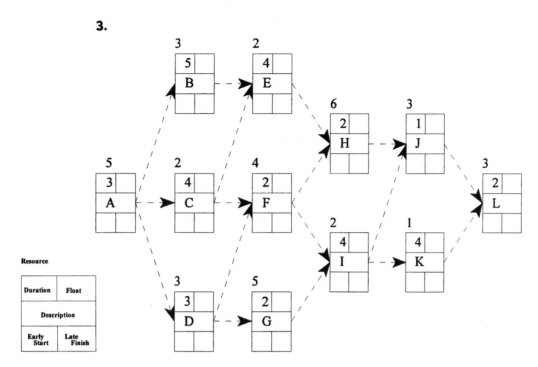

Figure 6–31   Exercise 3—Precedence Diagram with Labor Resource

4.

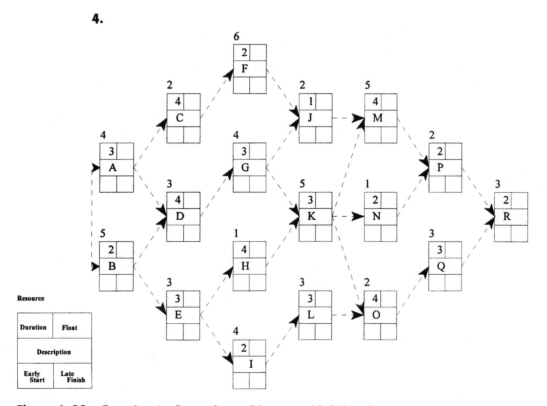

Figure 6–32   Exercise 4—Precedence Diagram with Labor Resource

5.

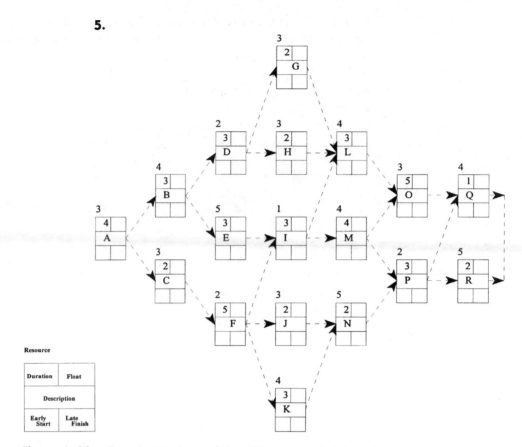

Figure 6–33   Exercise 5—Precedence Diagram with Labor Resource

6.

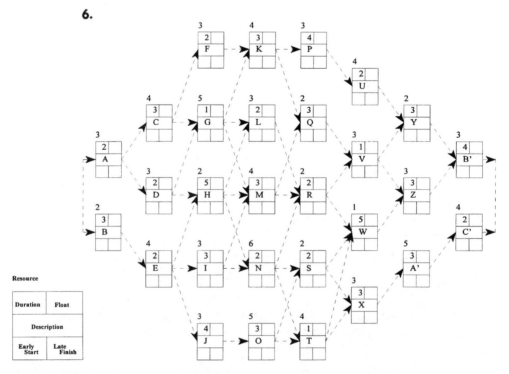

Figure 6–34   Exercise 6—Precedence Diagram with Labor Resource

7. Small Commercial Concrete Block Building—Resources

Prepare an on-screen bar chart for the small commercial concrete block building located in the Appendix. This exercise should include the following steps:

A. Prepare a list of labor resources required by activity.

B. Create the on-screen resource profiles.

C. Create the on-screen resource tables.

8. Large Commercial Building—Resources

Prepare an on-screen bar chart for the large commercial building located in the Appendix. This exercise should include the following steps:

A. Prepare a list of labor resources required by activity.

B. Create the on-screen resource profiles.

C. Create the on-screen resource tables.

7

Costs

Objectives

Upon completion of this chapter, you should be able to:

- Create cost accounts
- Assign activity costs
- Use cost profiles
- Use cost spreadsheets
- Use cumulative costs

NECESSITY OF CONTROLLING COSTS

The most important resource needed for building is money. Scheduling and controlling the expenditure of funds are critical to the building process.

Financing

The contractor must finance the project. Having the right amount of funds necessary at the right time is a tricky business. Ultimately, the owner pays for a project through a contractor, but the contractor must make sure that money is available to finance interim periods until the monthly payment from the owner is available. With a good estimate, accurate cash projections, and intelligent distribution of funds, the contractor can finance the project without having to borrow funds. If the contractor has to borrow funds, a potential source of profit is lost in interest payments to a banker.

Conflict. Assume a contractor's bid to an owner for a new project is $10 million. The contractor never has that much money in the bank available to completely finance the project. The contractor depends on progress payments to recoup costs as the project is being put in place. Every contractor wants to finance the project completely with the owner's money; every owner wants the contractor to have some money at financial risk in the project.

Cash Needs. The contractor has to be able to finance not only that particular project but other projects he or she is building, while having funds to finance the home office. Money has to be available to meet payroll costs each week for the craftspersons who work directly for the contractor. Money has to be available each month to pay for materials and supplies used at the job site and for project subcontractors. The contractor also must finance the construction equipment used at the job site, whether it is owned by the company or rented.

Timing the Expenditures. The contractor needs to maintain cash flow to finance projects without having to borrow funds or take cash from accounts where it is earning interest. The goal is to have each project stand on its own. This requires:

- Control of payment requests to owner
- Control of labor productivity and costs
- Control of material and supplies costs
- Control of subcontractors costs
- Control of overhead costs
- Control of payment of funds

Owner's Requirements

The owner is as concerned about controlling the expenditure of funds as he or she is of controlling time. Most project contracts call for the contractor to be paid monthly as the project progresses. The owner needs a way to ensure not only that the project is progressing toward a satisfactory completion but also that funds are being properly spent and that the contractor is not being overpaid according to the progress accomplished on the project.

Measure Progress. The primary reason for the owner requiring a schedule is to have a document to evaluate the progress of the contractor. Usually, within a short period of time after the contract is signed, the contractor must provide the owner with two documents. The first is the schedule and the second is the schedule of values. The schedule of values is a breakdown of cost by category or phase of work. When the contractor turns in a pay request at the end of the month, a judgment is made as to the percent completion of each category. Since this judgment is usually subjective, there is room for argument. Many owners now require an activity cost-loaded schedule. An activity cost-loaded schedule provides more detailed breakdown of cost. The process of cost-loading the activities and coming to an early agreement as to their value reduces potential conflict. There is less argument at each pay period as to percent complete, since the completion of a particular activity is much easier to judge than progress in broad phases of cost.

The purpose of this chapter is to address the planning of cash needs and expenditures and the timing of the expenditures so that the contractor is in a constant positive cash flow position.

Manual Calculation.   Table 7–1 shows the total cost input for each activity and the cost per day for each activity in the Sample Schedule. For Activity A1000, Clear Site, the cost per day is the Total Cost, $320, divided by the Duration, 2. The cost per day for this activity is $160. The total cost for the entire project is $4,172, or the summation of all activities.

Table 7–2 shows the costs according to a timescale and how the money is expected to be spent. The planned $320 expenditure for Activity A1000 is to be spent on the first ($160) and second days ($160) of the project, or Monday and Tuesday. According to our plan, $160 will be spent on the second day of the project, but we will have spent a cumulative total of $320 by the end of the day. This includes the $160 spent on day 1 and the $160 spent on day 2. So, we know how much will be spent at the end of each day toward the total cost of the project of $4,172. We also know that we will need $2,888 to finance the project through the end of the sixth day, Monday.

Activity		Duration	Budgeted Total Cost	Budgeted Cost/Day
A1000	Clear Site	2	$320	$160
A1010	Building Layout	1	$346	$346
A1020	Excavate Footings	2	$739	$370
A1030	Fabricate Ftg Rebar	1	$144	$144
A1040	Fabricate Ftg Formwork	2	$947	$473
A1050	Place Ftg Rebar & Formwork	2	$784	$392
A1060	Place Ftg Conc	1	$892	$892
	Total		$4,172	

Table 7–1   Sample Schedule—Cost/Day

INPUT COSTS

Cost Accounts.   A cost account structure enables the constructor to merge the estimating/cost accounting structure and budget with the schedule. *P3e*'s cost accounts are hierarchical, and they enable you to track activity costs according to your organization's specific cost account codes.

Activity		Duration	Day 1 Mon	Day 2 Tue	Day 3 Wed	Day 4 Thu	Day 5 Fri	Day 6 Mon	Day 7 Tue	Day 8 Wed
A1000	Clear Site	2	$160	$160						
A1010	Building Layout	1			$346					
A1020	Excavate Footings	2				$369.50	$369.50			
A1030	Fabricate Ftg Rebar	1				$144.00				
A1040	Fabricate Ftg Formwork	2				$473.50	$473.50			
A1050	Place Ftg Rebar & Formwork	2						$392.00	$392.00	
A1060	Place Ftg Conc	1								$892.00
	Cost/Day:		$160	$160	$346	$987.00	$843.00	$392.00	$392.00	$892.00
	Cumulative Cost/Day:		$160	$320	$666	$1,653.00	$2,496.00	$2,888.00	$3,280.00	$4,172.00

Table 7–2   Sample Schedule—Cumulative Cost

To create a cost account hierarchy, select **Cost Accounts** from the **Enterprise** main pull-down menu (Figure 7–1). The **Cost Accounts** dialog box will appear (Figure 7–2). Click the **Cost Account ID** column to display the cost accounts hierarchy. An outline icon in the **Cost Account ID** column label controls the hierarchy display. To add a new cost account, select the cost account from the **Cost Account ID** column immediately above and of the same hierarchy level as the cost account you want to add; then click the **Add** button (Figure 7–2). Then type in the new cost account ID, the new cost account name, and a brief cost account description.

For the example purposes of this book, only total cost per activity is input. A detailed cost account structure has not been inputted. So, for the purposes of the Sample Schedule, a new **Cost Account ID** of **Cost** is created (see Figure 7–2). **Total Cost** is the new **Cost Account Name** and **Total Cost** is also input as the **Cost Account Description.**

Now that this simple cost account structure has been inputted, the next step is input of cost by activity. With the **Gantt Chart** and **Activity**

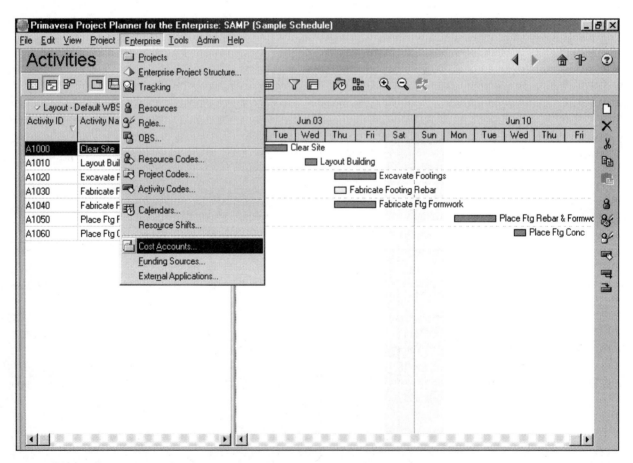

Figure 7–1 **Enterprise** Main Pull-Down Menu

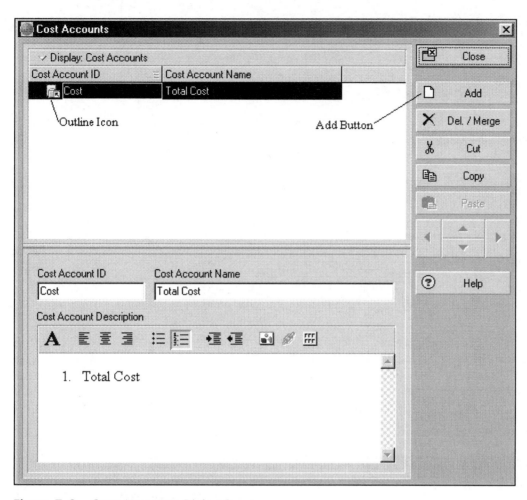

Figure 7–2   Cost Accounts Dialog Box

Details buttons selected, click on the **Expenses** tab (Figure 7–3). Click on Activity **A1000** to select it. Click the **Add** button, from the **Expenses** tab, and the **New Expense Item** line will appear (Figure 7–4). In the **Expense Item** field, **Total Cost** is entered (Figure 7–5). Next, click in the **Cost Account** field and the **Select Cost Account** dialog box will appear (Figure 7–5). The **Cost Account ID** of **Cost** is selected and the **Select** button is clicked. Note that, in Figure 7–6, **Cost** appears as the **Cost Account** selected.

The next step is to input the activity budgeted cost. From Table 7–1, the budgeted total cost for Activity **A1000** is $320. Enter the $320 in the **Budgeted Cost** field in the **Expenses** tab (Figure 7–6). To input the budgeted cost for Activity **A1010**, simply click on the activity to select it and follow the above steps. The budgeted costs for the sample schedule are input according to Table 7–1.

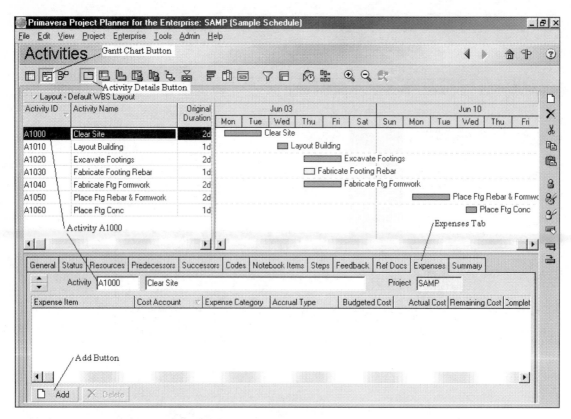

Figure 7–3   Activity Details Window—**Expenses** Tab

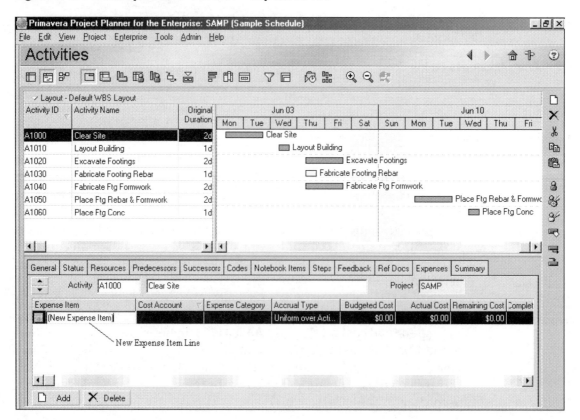

Figure 7–4   Expenses Tab—New Expense Item Line

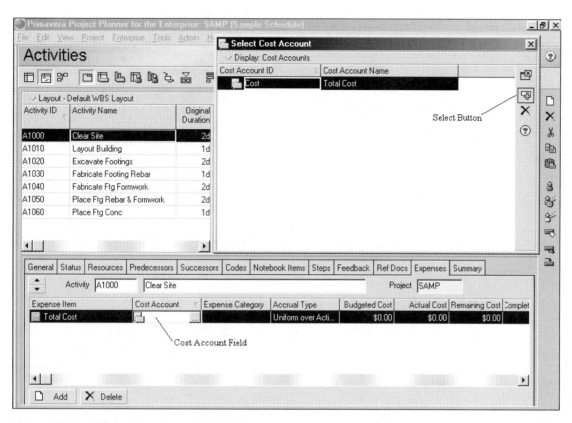

Figure 7–5   Select Cost Account Dialog Box

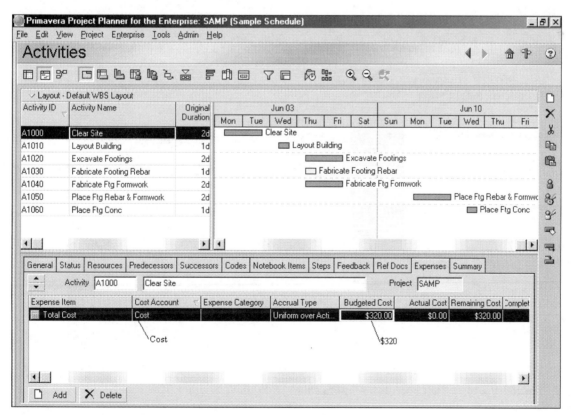

Figure 7–6   Expenses Tab—**Budgeted Cost**

COST SPREADSHEET

To view the **Cost Spreadsheet**, click on the **Activity Usage Spreadsheet** button from the Activity Toolbar. The last time the **Activity Usage Spreadsheet** was used in Chapter 6, it was configured to show resources. The configuration must now be changed to show cost. To change the configuration, click the right mouse button in the table portion of the **Activity Usage Spreadsheet**, and select <u>S</u>preadsheet Fields from the menu. The **Fields** dialog box will appear (Figure 7–7). Click the + button beside the **Time Interval** portion of the **Available Options** column to expose the options. Next expose the **Costs** options under **Time Interval**, and find the **Budgeted Total Cost** option. Then click the right pointing arrow button (Figure 7–7) to move the **Budgeted Total Cost** option to the **Selected Options** column.

Note that, in Figure 7–7, the **Budgeted Total Cost** option has been selected. Click the **OK** button of the **Fields** dialog box to accept the new configuration. Figure 7–8 is the result of the new cost configuration of the **Activity Usage Spreadsheet**. The **Activity Usage Spreadsheet** in Figure 7–8 matches Table 7–2, showing the total cost per day requirements for the Sample Schedule.

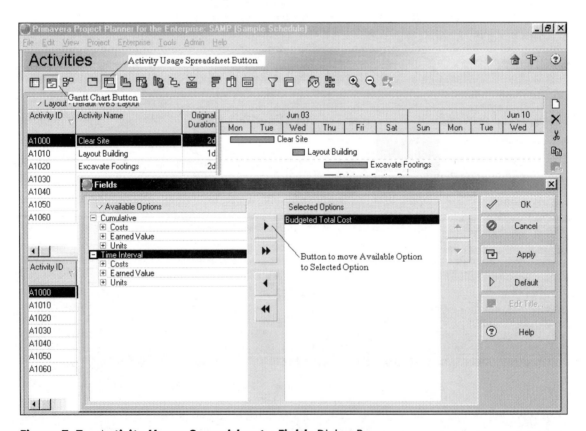

Figure 7–7   Activity Usage Spreadsheet—Fields Dialog Box

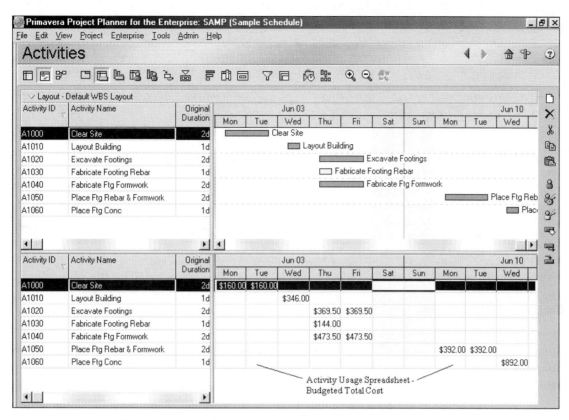

Figure 7–8   Activity Usage Spreadsheet—Budgeted Total Cost

ASSIGN OTHER COSTS

So far in this chapter, only **Budgeted Cost** from the **Expenses** tab of the **Activity Details** window has been used (Figure 7–6). This approach is fine for pricing only total cost per activity, but *P3e*'s resource pricing capabilities per individual resource have not been used. The next few figures are shown as an example of the changes within *P3e* when individual costs are added at the resource level. To observe *P3e*'s resource pricing capabilities, click on **Resources** from the **Enterprise** main pull-down menu and the **Resources** screen will appear (Figure 7–9).

The **CARP 1** resource is selected in Figure 7–9. When this resource was originally defined, the **Price/Time** field was set at $0.00. The **Price/Time** fields for all the other resources input for the Sample Schedule were also set at $0.00 (Figure 6–20). The **Activity Usage Spreadsheet** in Figure 7–8 is generated using the $0.00 value for the **Price/Time** rate for the resources as input in the last chapter. The **Price/Time** column for **CARP 1** has been changed from $0.00 to $15.00 in Figure 7–9.

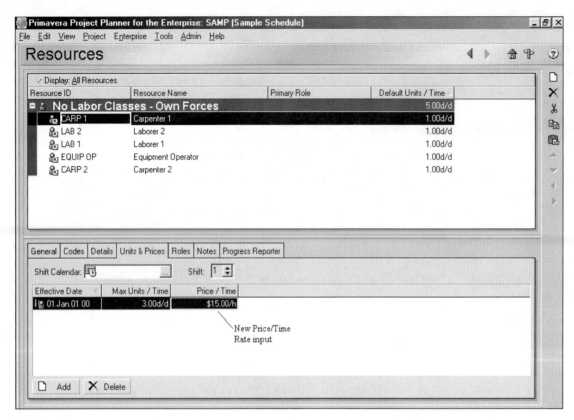

Figure 7–9   Resources Screen

The impact of changing the **Price/Time** field of the **Resources** screen can be seen by returning to the **Resource Usage Spreadsheet** (Figure 7–10). The configuration must now be changed to show cost (resources were shown the last time we used the spreadsheet). To change the configuration, click the right mouse button in the table portion of the **Resource Usage Spreadsheet**, and select **Spreadsheet Fields** from the menu. The **Fields** dialog box will appear (Figure 7–10). Click the + button beside the **Time Interval** portion of the **Available Options** column to expose the options. Next expose the **Costs** options under **Time Interval**, and find the **Budgeted Cost** option. Then click the right pointing arrow button to move the **Budgeted Cost** option to the **Selected Options** column.

Note that, in Figure 7–10, the **Budgeted Cost** option has been selected. Click the **OK** button of the **Fields** dialog box to accept the new configuration. Figure 7–11 is the result of the new cost configuration of the **Resource Usage Spreadsheet**. The **CARP 1** resource is selected; therefore, the spreadsheet shows all activities where **CARP 1** is specified.

To evaluate the overall activity impact of changing the **Price/Time** field of the **Resources** screen, click on the **Activity Usage Spreadsheet**. Compare the **Activity Usage Spreadsheet** in Figure 7–12, after the **Price/Time**

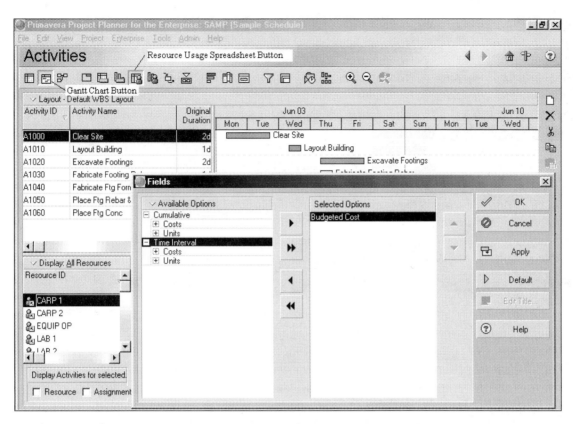

Figure 7–10   Resource Usage Spreadsheet—Fields Dialog Box

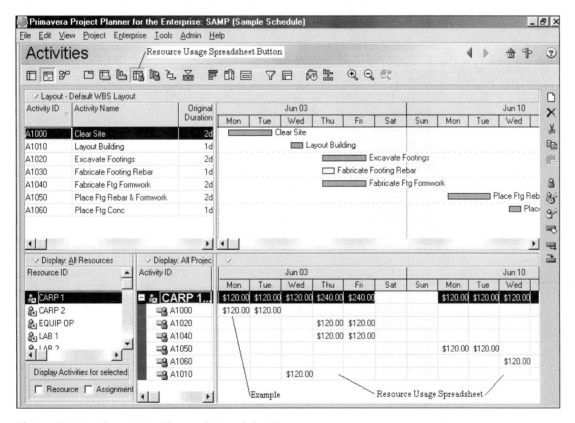

Figure 7–11   Resource Usage Spreadsheet

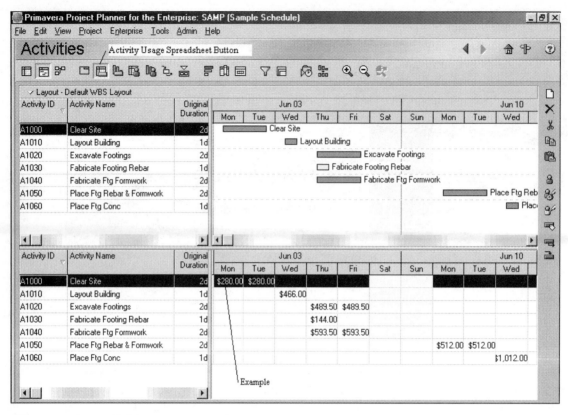

Figure 7–12   Activity Usage Spreadsheet

change for **CAR** has been added, to that in Figure 7–8, before the change. The $280.00 value appearing in the Monday column of the **Activity Usage Spreadsheet** (Figure 7–12) is composed of $160.00 (Figure 7–8) plus $120.00 (Figure 7–11). The $160.00 value from Figure 7–8 is generated from the **Budgeted Cost** from the **Expenses** tab of the **Activity Details** window (Figure 7–6). The $120 value from Figure 7–11 is generated from the **Price/Time** field of the **Resources** screen (Figure 7–9). The ability to price individual resources at a unit price and let *P3e* extend the prices on the spreadsheets provides a very beneficial cost analysis tool.

COST PROFILES

To view an overall project cost profile, click on the **Activity Usage Profile** button (Figure 7–13) . The **Activity Usage Profile** will appear at the bottom of the screen (Figure 7–14).

Configurations.   If when you view the **Activity Usage Profile**, no usable information is included, you must reconfigure the table graphic. Double click on the **Activity Usage Profile** and the **Activity Usage**

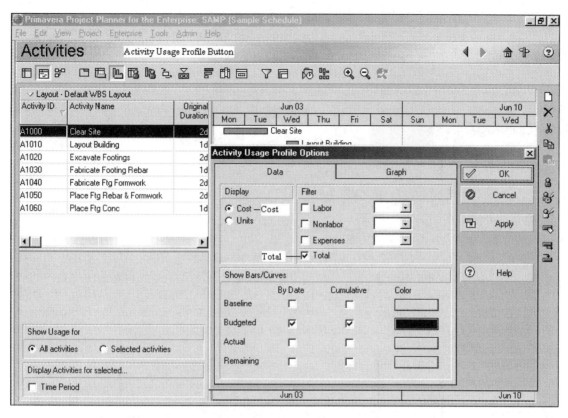

Figure 7–13   Activity Usage Profile Options Dialog Box

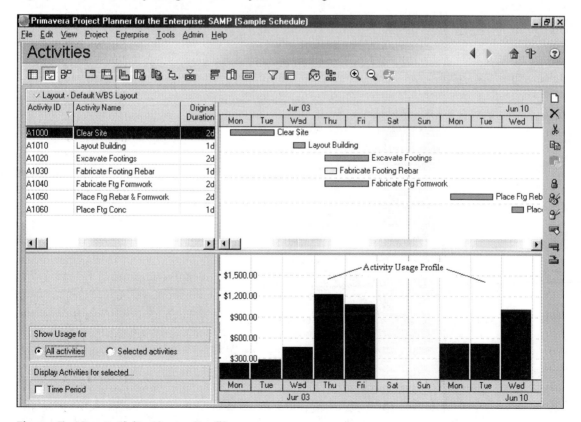

Figure 7–14   Activity Usage Profile

Profile Options dialog box will appear (Figure 7–13). Under the **Display** field, click on the **Cost** option. You may recall that in Chapter 6 we viewed resources and selected units using this same dialog box. Note that, in Figure 7–13, the checkbox **Total** is checked under the **Filter** portion of the **Activity Usage Profile Options** dialog box. This means that the **Labor** and **Expenses** cost as input earlier in this chapter will appear in the **Activity Usage Profile**. The $280.00 value appearing in the Monday column of the **Activity Usage Profile** (Figure 7–14) is composed of $160.00 (Figure 7–8, **Expenses** portion) plus $120.00 (Figure 7–11, **Labor** portion). The $160.00 value from Figure 7–8 is generated from the **Budgeted Cost** from the **Expenses** tab of the **Activity Details** window (Figure 7–6). The $120 value from Figure 7–11 is generated from the **Price/Time** field of the **Resources** screen (Figure 7–9).

Instead of looking at the overall project cost profile (**Activity Usage Spreadsheet**), you may want to analyze individual resources. To select an individual resource for analysis, select the **Resource Usage Profile** button (Figure 7–15). Note that the **CARP 1** resource has been selected in Figure 7–15. Since the **Resource Usage Profile** deals with resources only,

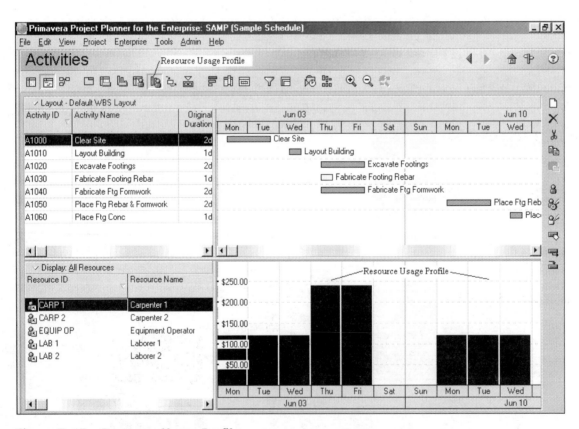

Figure 7–15 Resource Usage Profile

the $120 value generated from the **Price/Time** field of the **Resources** screen (Figure 7–9) appears in the **Resource Usage Profile** for the Monday column. The **Expenses** value of $160.00 as generated from the **Budgeted Cost** from the **Expenses** tab of the **Activity Details** window (Figure 7–6) does not appear.

CUMULATIVE COSTS

Cumulative Curve.   From the Activity Usage Profile Options dialog box (Figure 7–16), click on the **Cumulative** checkbox. The cost profile, displaying cost per day, appears at the bottom of the screen (Figure 7–17).

Area under the Curve.   The area under the curve represents the cumulative amount of money spent on the project to date. The vertical scale on the right of the cost profile represents the total cumulative cost

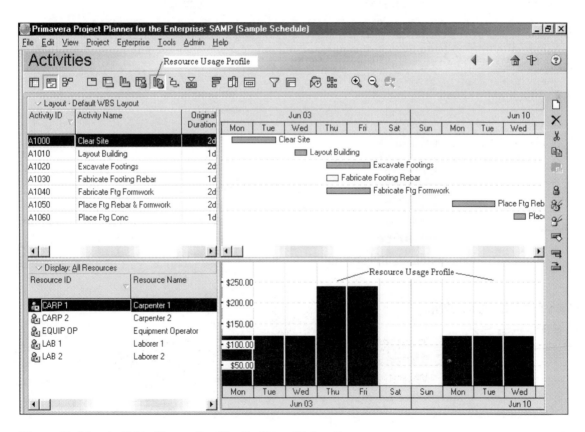

Figure 7–16   Activity Usage Profile Options Dialog Box

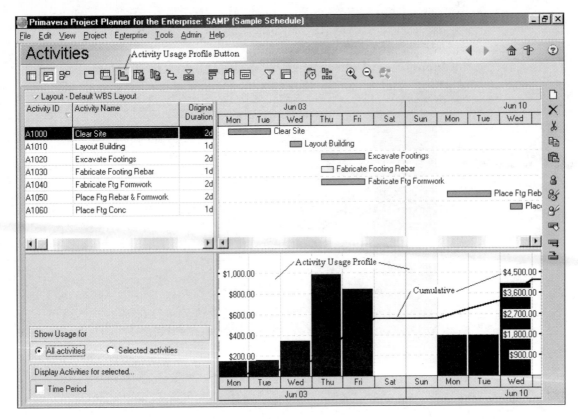

Figure 7–17   Activity Usage Profile—Cumulative Curve

for the project. The total monies anticipated to be spent (cash flow) for the project is $4,172 (Figure 7–17). No money is spent until the beginning of the day on Monday, June 4. The last activity for which money is expended takes place on Wednesday, June 13. The anticipated monies to be spent on the project will be spent between June 4 and 13. The height of the curve shows the monies to be spent as of any particular day.

EXAMPLE PROBLEM

Table 7–3 is an Activity List with expenses for a house put together as an example for student use (see the wood-framed house drawings in the

Activity ID	Activity Description	Costs
A1000	Clear Site	$1,280
A1010	Building Layout	$386
A1020	Form/Pour Footings	$1,174
A1030	Pier Masonry	$967
A1040	Wood Floor System	$4,181
A1050	Rough Framing Walls	$3,323
A1060	Rough Framing Roof	$3,468
A1070	Doors and Windows	$3,995
A1080	Ext Wall Board	$736
A1090	Ext Wall Insulation	$385
A1100	Rough Plumbing	$750
A1110	Rough HVAC	$1,168
A1120	Rough Elect	$940
A1130	Shingles	$1,091
A1140	Ext Siding	$1,710
A1150	Ext Finish Carpentry	$736
A1160	Hang Drywall	$1,844
A1170	Finish Drywall	$790
A1180	Cabinets	$1,618
A1190	Ext Paint	$525
A1200	Int Finish Carpentry	$1,472
A1210	Int Paint	$4,725
A1220	Finish Plumbing	$3,000
A1230	Finish HVAC	$3,506
A1240	Finish Elect	$1,410
A1250	Flooring	$1,583
A1260	Grading & Landscaping	$600
A1270	Punch List	

Table 7–3   Activity List with Costs—Wood-Framed House

Appendix). Figures 7–18 and 7–19 provide additional resource information. The **Activity Usage Spreadsheet** for Cost (Figure 7–18) and the **Activity Usage Profile** for Cost (Figure 7–19) were prepared using the tabular list of 28 activities (activity ID, description, and total activity costs).

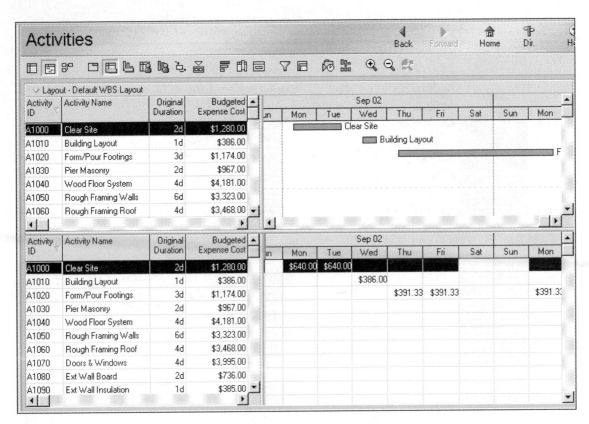

Figure 7–18 Activity Usage Spreadsheet for Cost—Wood-Framed House

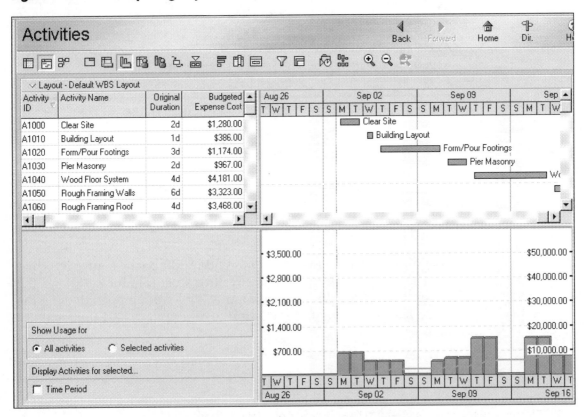

Figure 7–19 Activity Usage Profile for Cost—Wood-Framed House

EXERCISES

Complete a manual cost/day and cumulative cost table similar to Tables 7–1 and 7–2 for Exercises 1 through 6. Also plot the cumulative curves.

1.

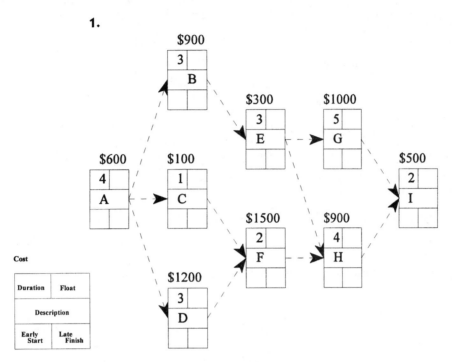

Figure 7–20   Exercise 1—Cost Table and Cumulative Curve

2.

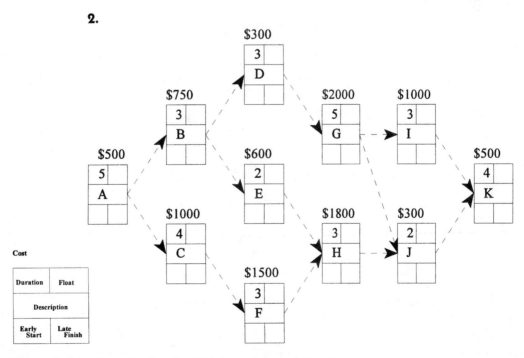

Figure 7–21   Exercise 2—Cost Table and Cumulative Curve

3.

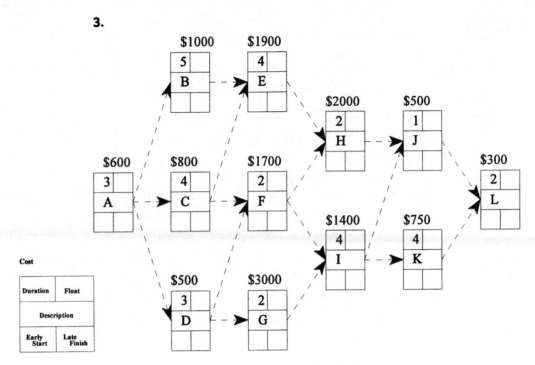

Figure 7–22 Exercise 3—Cost Table and Cumulative Curve

4.

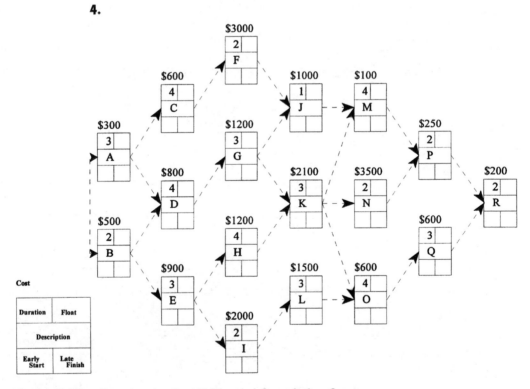

Figure 7–23 Exercise 4—Cost Table and Cumulative Curve

5.

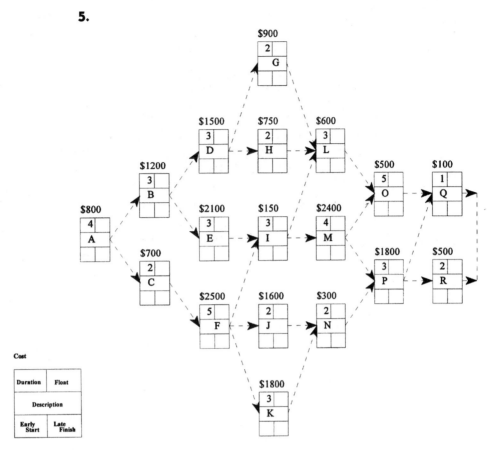

Figure 7–24   Exercise 5—Cost Table and Cumulative Curve

6.

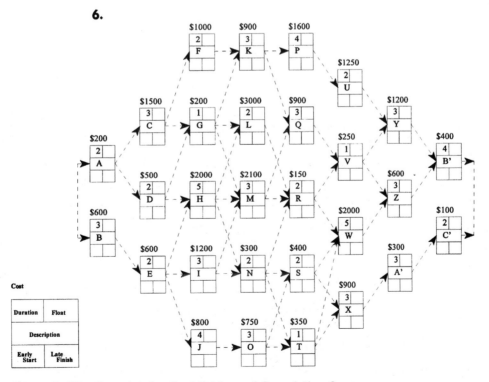

Figure 7–25   Exercise 6—Cost Table and Cumulative Curve

7. **Small Commercial Concrete Block Building—Costs**
   Prepare an on-screen cost profile and cost table for the small commercial concrete block building located in the Appendix. This exercise should include the following steps:
   A. Assign costs per activity by type of cost.
   B. Create the on-screen cost profile.
   C. Create the on-screen cost table.

8. **Large Commercial Building—Costs**
   Prepare an on-screen cost profile and cost table bar chart for the large commercial building located in the Appendix. This exercise should include the following steps:
   A. Assign costs per activity by type of cost.
   B. Create the on-screen cost profile.
   C. Create the on-screen cost table.

Gantt Charts—Print Preview

Objectives

Upon completion of this chapter, you should be able to:

• Modify a Gantt chart print
• Print a Gantt chart

NECESSITY FOR GOOD PRESENTATIONS

Obviously, putting a schedule together at the beginning of a project is valuable. The process forces you to plan, organize, sequence, and show the interrelationships among different activities of the project. This information is useless, however, unless it is communicated to all relevant parties on the project. Good scheduling involves disseminating information to and getting feedback from all key parties regarding the original plan, updates, and progress toward completion of the project. At present, the easiest and most efficient way to disseminate the computerized *P3e* schedule is through hard-copy (paper) graphical or tabular reports. E-mail with attached files, or using of the Internet to transfer graphical information, is moving construction companies toward a paperless office of schedule information.

GRAPHIC REPORTS

A graphical report is a printout or plot depicting information about the schedule. A plotter produces large drawings (Table 8–1). Plotters offer multiple colors, line types, line widths, and shading capabilities for producing professional presentations. Graphical printers are usually of the laser jet or ink jet variety. Multiple colors are an option with these printers and the paper is usually letter size (8.5" × 11") or legal size (8.5" × 14"). Whether you choose a printer or a plotter depends primarily on the size of the hard copy you wish to produce.

U.S.	Letter	8 1/2 × 11 in
	Legal	8 1/2 × 14 in
ANSI	B	11 × 17 in
	C	17 × 22 in
	D	22 × 34 in
	E	34 × 44 in
Architectural	A	9 × 12 in
	B	12 × 18 in
	C	18 × 24 in
	D	24 × 36 in
	E	36 × 48 in
ISO	A4	210 × 297 mm
	A3	297 × 420 mm
	A2	420 × 594 mm
	A1	594 × 841 mm
	A0	849 × 1189 mm

Table 8–1   Printer/Plotter Paper Sizes

GANTT CHART PRESENTATION

As mentioned earlier, the primary advantage of the Gantt chart is its overall simplicity. It is easy to read and interpret, and therefore it can be an effective communications tool. The primary disadvantage is that it does not normally show interrelationships among project activities. By selecting **Bars** from the **View** main pull-down menu, then selecting the **Options** button, and then the **Show Relationships** checkbox, you can view the logic interrelationships on-screen. If an activity is delayed and the relationship lines are not shown, it is difficult to determine the impact on the rest of the schedule. The Gantt chart is undoubtedly the most commonly used print form of *P3e*.

A hard copy of the *P3e* bar chart can be produced in one of two ways. It can be accessed either from the **File** or the **Tools** main pull-down menus through **Reports**.

The **File** option (Figure 8–1) gives essentially a copy of the on-screen Gantt chart. As you can see from Figure 8–1, there are three print functions available: **Page Setup**, **Print Preview** and **Print**.

PRINT PREVIEW OPTIONS

Toolbar.   The **Print Preview** toolbar options are customizable by the user. What we show in this chapter are the *P3e* defaults. Click on the

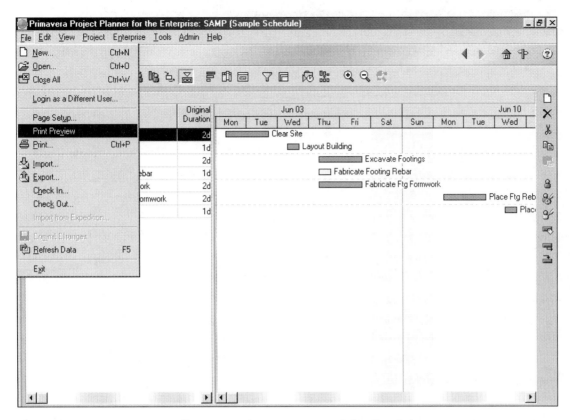

Figure 8–1   Gantt Chart—**File** Main Pull-Down Menu—**Print Preview**

Print Preview option and the **Print Preview** screen will appear (Figure 8–2). It is an on-screen image of the hard-copy print of the *P3e* on-screen image. You can manipulate the on-screen image of the print using the toolbar at the top of the screen.

PRINT PREVIEW BUTTON

The **Print Preview** button (Figure 8–2) can be used to close print preview and return you to the on-screen active image.

PAGE VIEW BUTTONS

The **Page View** buttons (Figure 8–2) are arrows for selecting page views. If the schedule print is composed of multiple pages, use the horizontal arrows to scroll the pages horizontally and the vertical arrows to scroll pages vertically. In Figure 8–2, four of the six arrows are "grayed out," meaning that the hard-copy print of the Sample Schedule is composed of

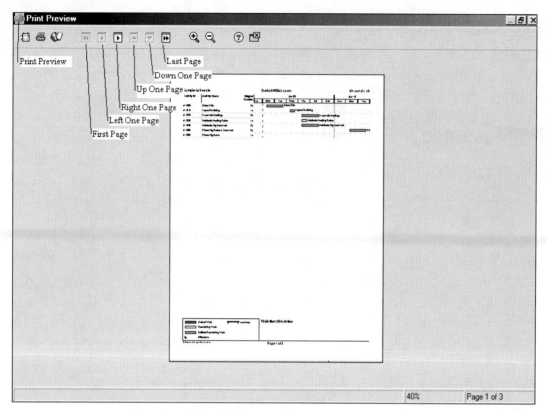

Figure 8–2   Gantt Chart—**Print Preview**—Page Views

three horizontal pages. Click on the **Right One Page** button to view the second horizontal page.

ZOOM BUTTONS

Zoom In.   The **Zoom In** button of the **Print Preview** screen (Figure 8–3) is identified by a magnifying glass with a plus sign inside. When the cursor is outside the boundaries of the reproduction of the hard-copy page, it appears as an arrow (normal appearance). When the cursor is inside the boundaries of the on-screen hard copy page, it appears as a magnifying glass. Move the cursor (magnifying glass) to the location to be zoomed (or magnified); then click the mouse. *P3e* automatically sizes the print. When **Zoom In** is selected, *P3e* zooms in to the next standard zoom level (60%, 90%, 135%, or 200%), making the image larger. Figure 8–3 is a copy of Figure 8–2 zoomed in once (90%) with the mouse cursor pointed to the upper section of the screen (the section of the screen that we want zoomed in on).

Zoom Out.   The **Zoom Out** button (Figure 8–3) is identified by a magnifying glass with a minus sign inside. By clicking on this button, *P3e*

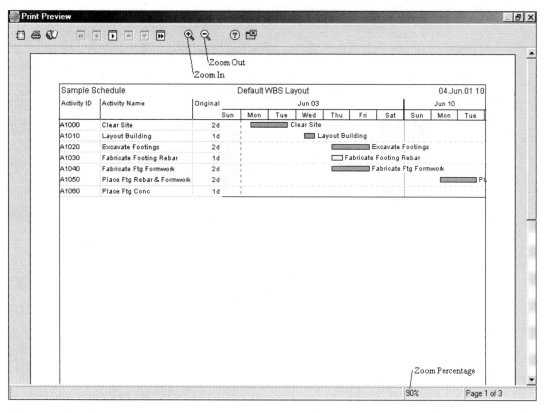

Figure 8–3   Zoom Buttons

automatically zooms out to its next standard zoom level (making the image smaller).

Sometimes the standard zoom levels used in **Zoom In** and **Zoom Out** do not meet your needs for custom fitting a hard-copy print. Print **Scaling** options used to meet these custom size requirements are available under the **Page Setup** button options of **Print Preview**. These options will be covered later in this chapter.

PRINT BUTTON

The **Print** button (see Figure 8–4) is used to actually execute the hard-copy print of the **Print Preview** screen. Click on the **Print** button to produce the **Print** dialog box (Figure 8–5).

Print Range.   This function is used to select a limited number of pages rather than printing the entire schedule. Suppose there is a schedule that has a hard copy of six pages, three pages horizontally (numbers) and two

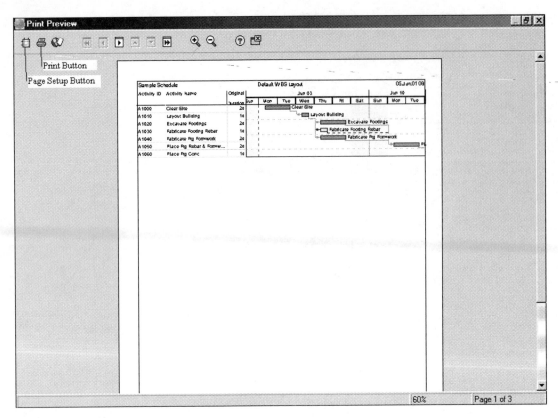

Figure 8–4   Print and **Page Setup** Buttons

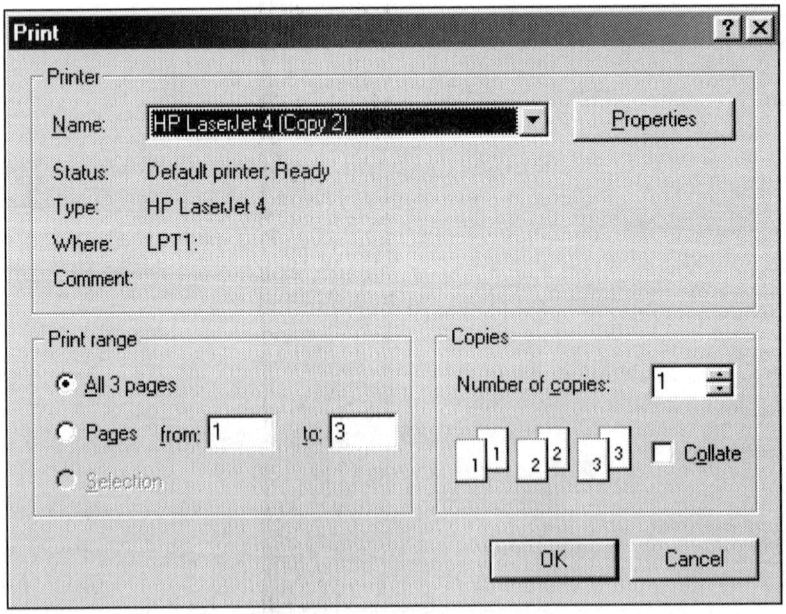

Figure 8–5   Print Dialog Box

pages vertically (alpha characters). The *P3e* numbering scheme for this combination is:

1A   1B   1C
2A   2B   2C

By understanding this numbering scheme, you can avoid having to make hard-copy prints for pages that are not necessary.

Copies.   This option is used if you want multiple copies of each page of the hard-copy print.

PAGE SETUP BUTTON

See Figure 8–4 for the location of the **Page Setup** button. This button is used to modify the page setup of the hard-copy print of the **Print Preview** screen. Page setup of the hard copy can be accessed through **Print Preview** or directly from the **File** main pull-down menu (Figure 8–1).

Clicking the **Page Setup** button produces the **Page Setup** dialog box (Figure 8–6). The **Page Setup** dialog box has six tabs with options for modifying the **Print Preview** and therefore hard-copy prints. The **Page Setup** tabs are **Page**, **Margins**, **Header**, **Footer**, **Legend**, and **Options**. A discussion of each of these tabs follows.

Page Tab.   The first of the six **Page Setup** dialog box tabs is the **Page** tab (Figure 8–6). The **Page** tab provides the following fields for modifying the hard-copy print:

> **Orientation**   The Orientation field lets you choose between a **Portrait** and **Landscape** orientation. Figure 8–2 is an example of a **Portrait** orientation. In Figure 8–7, the **Page** tab was changed to produce a **Landscape** orientation.
>
> **Scaling**   The **Scaling** field enables you to customize the size of the *P3e* schedule within the **Print Preview** and therefore hard-copy prints. The **Zoom In** and **Zoom Out** buttons available in the **Print Preview** screen (Figure 8–3) have preset zoom percentages options. But, with the **Adjust to ____% normal size** field, *P3e* lets you pick custom zoom percentages. In Figure 8–7, a 60% zoom size was selected for scaling. The other **Scaling** field options are the **Fit to:** fields. The **page(s) wide by** and **tall** fields let you specify how many horizontal and vertical pages you want in the **Print Preview** and therefore hard-copy prints. *P3e* will automatically choose the scaling percentage to produce the desired result.

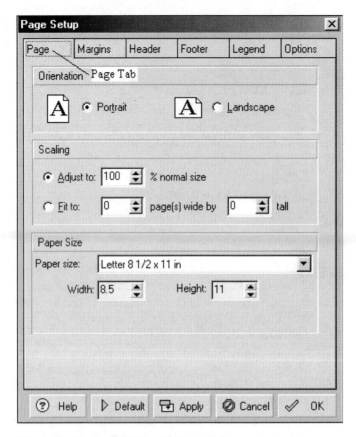

Figure 8–6   Page Setup—Page Tab

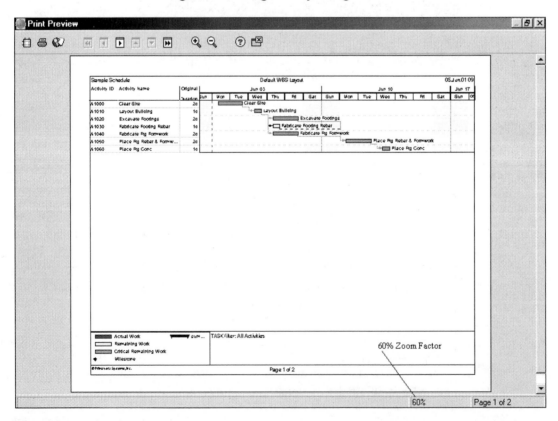

Figure 8–7   Landscape Format

Paper Size.   The down arrow available under the **Paper Size** field enables you to choose among the paper size options available for the printer that you have specified.

Margins Tab.   The second of the six **Page Setup** dialog box tabs is the **Margins** tab (Figure 8–8). The **Margins** tab provides you with the ability to change the margins of hard-copy prints from the *P3e* defaults that appear in Figure 8–8. Figure 8–9 shows the margins changed from the *P3e* defaults to 1.5 for all four margins.

Header Tab.   The third of the six **Page Setup** dialog box tabs is the **Header** tab (Figure 8–10). The **Header** tab provides you with the ability to change the information displayed in the header of the **Print Preview** and hard-copy *P3e* prints. Note the location of the header in the **Print Preview** in Figure 8–10. The **Header Sample** field, at the top of the **Header** tab, shows the actual information, as it will appear on the header. The **Header Sample** field combines the **Left**, **Center**, and **Right** header sections as defined in the bottom half of the **Header** tab.

Note in Figure 8–10 that the **Left** section of the header is selected, and **[project_name]** is the information selected for display. **Sample Schedule** is the project name given to our example project when it was created.

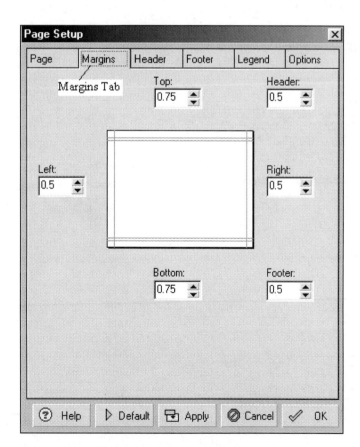

Figure 8–8   Page Setup—Margins Tab

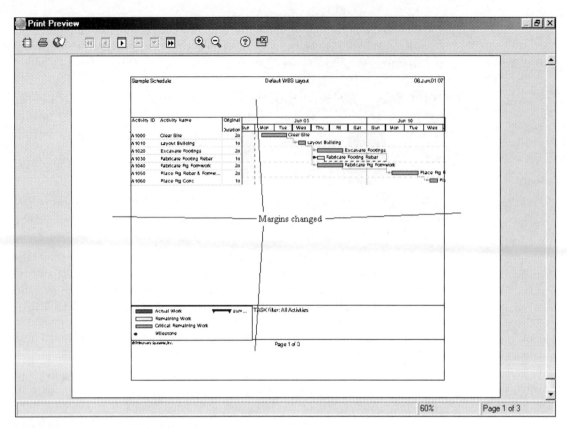

Figure 8-9   Print Preview—Margins Changed

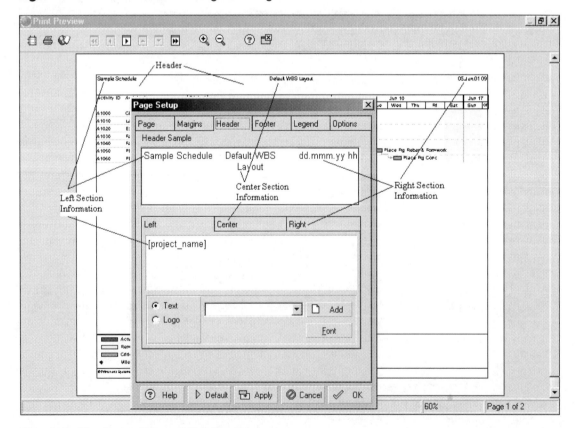

Figure 8-10   Page Setup—**Header** Tab

Observe that **Sample Schedule** appears in the **Sample Header** field of the **Header** tab and on the left side of the **Print Preview**. To place other information in the **Left** header section, other than **[project_name]**, place the mouse cursor where you want the new field to be located (beside, below, or replace the current field). Then click on the down arrow field to expose the drop-down menu of possible additional information options. Select the information option desired. Then click the **Add** button. Click the **Apply** button to cause the new information to appear in the **Header Sample** field of the **Header** tab and on the left side of the **Print Preview**.

Another option for including header information is shown in Figure 8–11. Instead of choosing a predefined information option from *P3e*, you can insert custom text. To obtain the **Print Preview** shown in Figure 8–11, the following changes are made. In the **Left** header section of the **Header** tab, the mouse cursor is placed after the **[project_name]** information field and the keyboard **Enter** key is struck to produce a second line of information. The new custom text, **Student Constructors**, is entered. The **Apply** button is then clicked to cause the new information to appear in the **Header Sample** field of the **Header** tab and on the left side of the **Print Preview**.

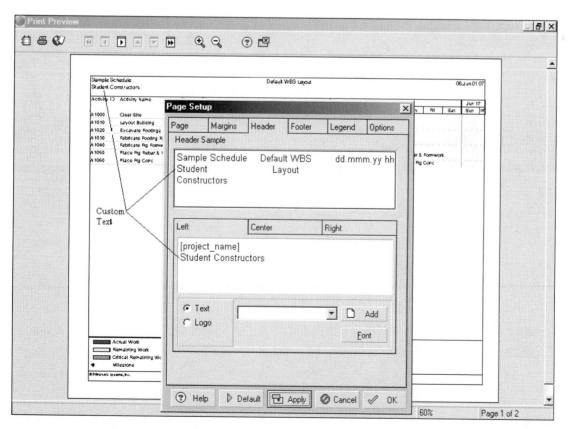

Figure 8–11   Header Tab—Custom Text

Footer Tab.   The fourth of the six **Page Setup** dialog box tabs is the **Footer** tab (Figure 8–12). The **Footer** tab provides you with the ability to change the information displayed in the footer of hard-copy *P3e* prints. The appearance and operation of this tab is identical to those of the **Header** tab. Note the location of the footer in the **Print Preview** in Figure 8–12.

As an example of using the available *P3e* information fields, new information is placed in the right footer field. In Figure 8–13, the **Right** tab has been selected and the down arrow has been clicked to expose the available information menu. The selection, **Filter List**, is clicked. Then the **Add** button is clicked to place the new selection on the **Right** tab of the **Footer** tab of the **Page Setup** dialog box (Figure 8–14). When the **Apply** button is clicked, the new information, **TASK filter: All Activities**, appears on the **Print Preview** (Figure 8–14).

Legend Tab.   The fifth of the six **Page Setup** dialog box tabs is the **Legend** tab (Figure 8–15). The **Legend** tab provides you with the ability to change the information displayed in the legend of hard-copy *P3e* prints. The appearance and operation of this tab is identical to the **Header** and **Footer** tabs. Note the location of the legend in the **Print Preview** in Figure 8–15.

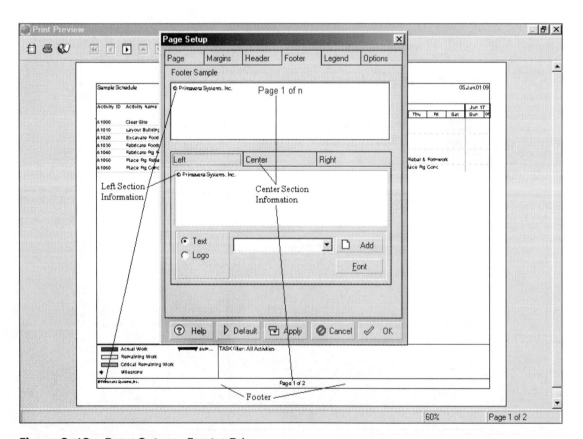

Figure 8–12   Page Setup—Footer Tab

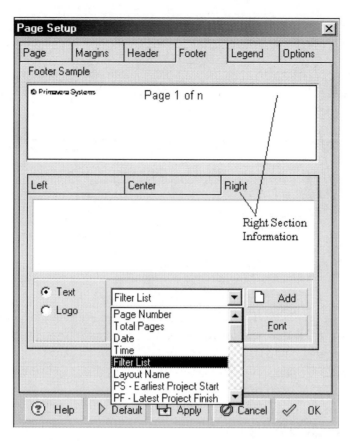

Figure 8–13   Information Options Available Field

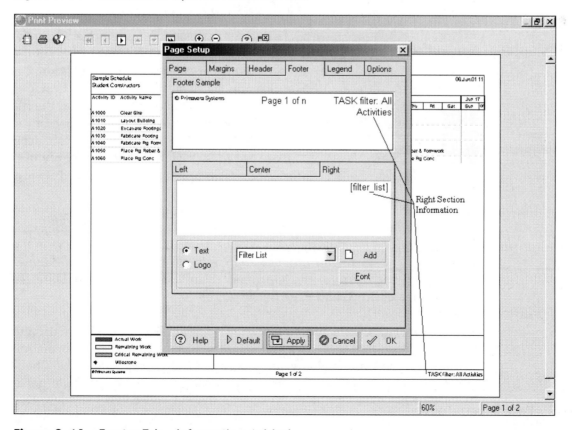

Figure 8–14   Footer Tab—Information Added

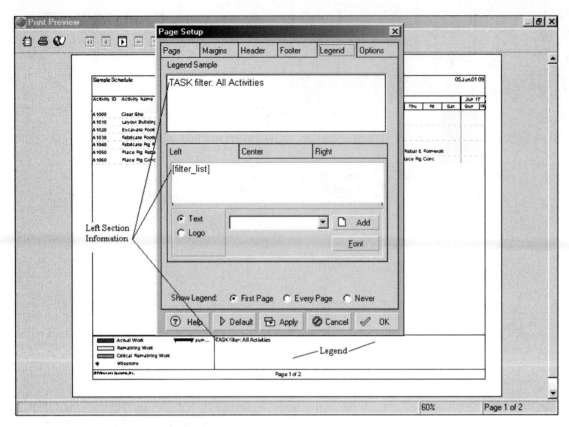

Figure 8–15 Page Setup—Legend Tab

Note that in Figure 8–16 the **[filter_list]** information is removed from the **Left** tab. Next, a company logo is to be placed in the right section of the legend. With the **Right** tab selected for location, the **Logo** checkbox is selected to place a company logo in the legend of the **Print Preview**. When the **Select Logo Filename** button is clicked to browse, find the file location and select the logo file. The logo file format must be one of the following: BMP files (*.BMP), Windows Metafiles (*.WMF), Enhanced Windows Metafiles (*.EMF), or Icon files (*.ICO). When the **Apply** button is clicked, the new information logo appears on the **Print Preview** (Figure 8–16).

Options Tab.   The sixth and last of the six **Page Setup** dialog box tabs is the **Options** tab (Figure 8–17). The **Options** tab provides you with the ability to change the information displayed on **Print Preview** and therefore hard-copy prints. The **Options** tab includes **From Start Date, To Finish Date, Print,** and **Page Setting** choices.

The **From Start Date** and **To Finish Date** fields (Figure 8–17) are used to enter custom start and finish dates for the time length of the information displayed on **Print Preview** and therefore hard-copy prints. If the schedule produced was for 12 months, you might want to distribute hard-copy prints for only the first two months. So you could use these two

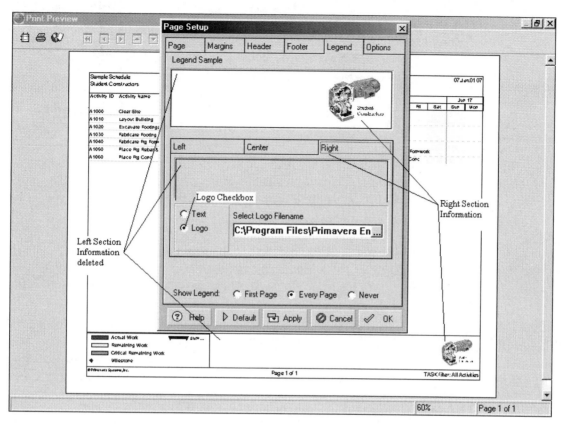

Figure 8–16   Legend Tab—Logo

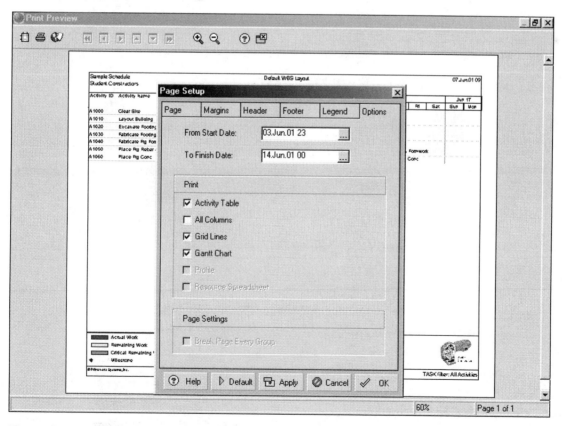

Figure 8–17   Page Setup—Options Tab

fields to customize the time length of the hard-copy prints. The **From Start Date** and **To Finish Date** fields relate only to the customized time length of the hard-copy prints. They have no bearing on the internal *P3e* forward pass, backward pass, or float calculations.

The **Print** field options include **Activity Table**, **All Columns**, **Grid Lines**, **Gantt Chart**, **Profile**, and **Resource Spreadsheet**.

Activity Table. Mark the **Activity Table** checkbox (Figure 8–18) to include the activity table in the **Print Preview** and therefore hard-copy prints. Compare Figure 8–9, with the activity table, to Figure 8–18, without the activity table.

All Columns. Mark the **All Columns** checkbox (Figure 8–19) to include the all columns available (both visible and hidden) in the on-screen Gantt chart in the **Print Preview**. Compare Figure 8–9, with only the visible on-screen Gantt chart columns appearing, to Figure 8–19, with all columns appearing. To change column locations or available columns, you must use the **Columns** dialog box. The **Columns** dialog box can be accessed either from the **Columns** button from the **Activity Toolbar** or by selecting **Columns** from the **View** main pull-down menu.

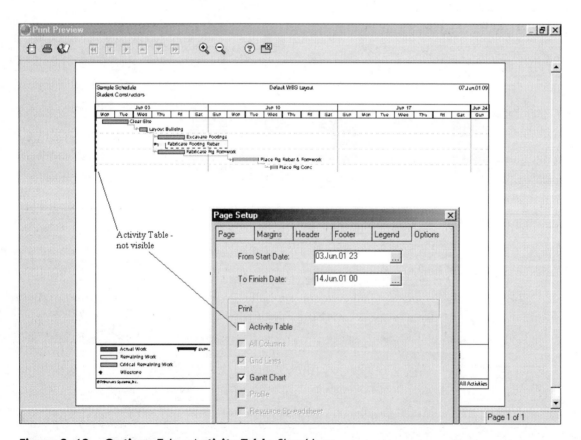

Figure 8–18   Options Tab—**Activity Table** Checkbox

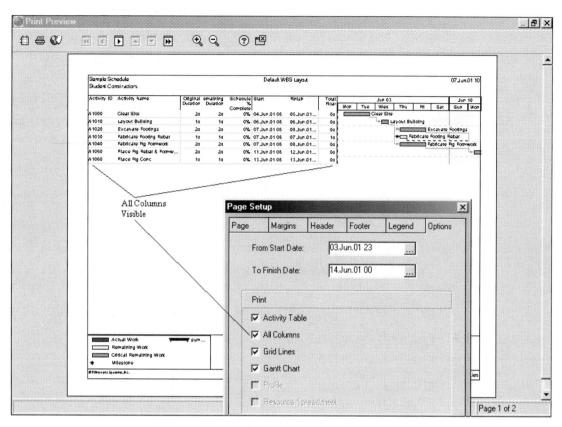

Figure 8–19   Options Tab—**All Columns** Checkbox

Grid Lines.   Mark the **Grid Lines** checkbox (Figure 8–20) to include the grid lines in the **Print Preview**. Compare Figure 8–20, without grid lines, to Figure 8–19, with grid lines.

Gantt Chart.   Mark the **Gantt Chart** checkbox (Figure 8–21) to include the Gantt chart in the **Print Preview**. Compare Figure 8–21, without the Gantt chart, to Figure 8–20, with the Gantt chart.

Profile.   Note that the **Profile** checkbox in Figure 8–21 is nonactive (grayed-out). The reason for it being nonactive is that, as stated earlier in the chapter, the **Print Preview** image is directly tied to the active on-screen image. To make this checkbox functional, the on-screen image in Figure 8–22 is changed for the **Gantt Chart** screen to include the **Activity Usage Profile** (bottom of the screen). Note that with this change, the **Profile** checkbox in Figure 8–23 is now visible.

Mark the **Profile** checkbox (Figure 8–23) to include the profile in the **Print Preview**. Compare Figure 8–23, with the profile, to Figure 8–9, without the profile.

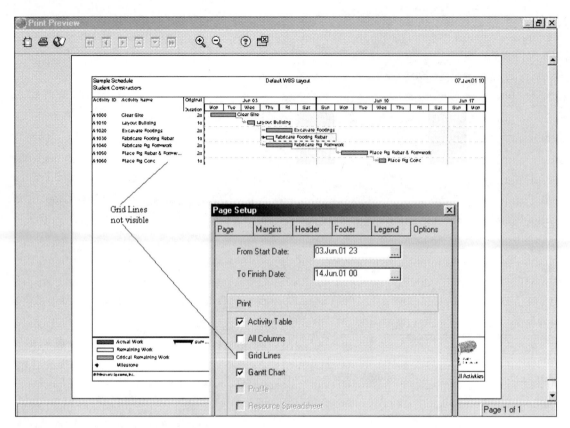

Figure 8–20 Options Tab—**Grid Lines** Checkbox

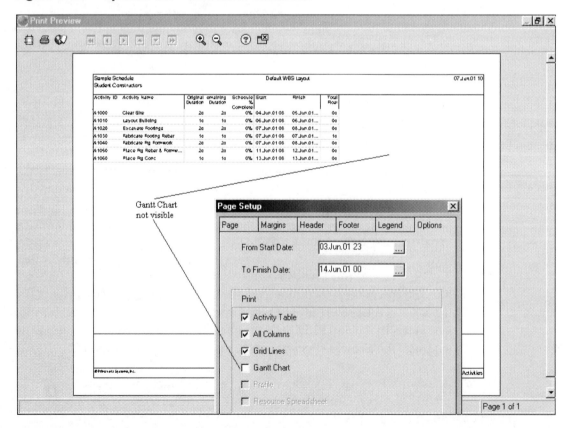

Figure 8–21 Options Tab—**Gantt Chart** Checkbox

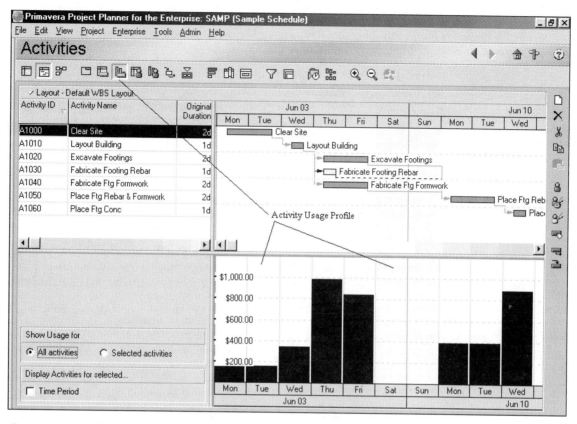

Figure 8–22   Gantt Chart Screen—**Activity Usage Profile**—Costs

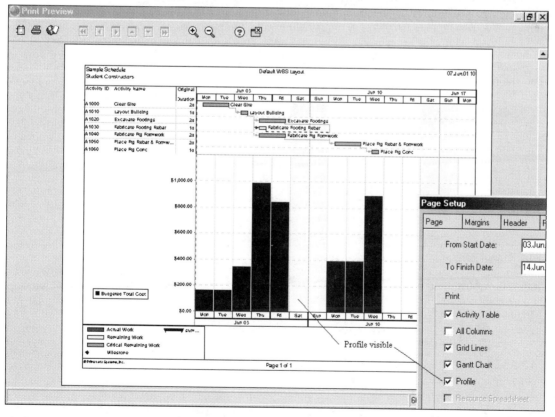

Figure 8–23   Options Tab—**Profile**

Resource Spreadsheet.   Note that the **Resource Spreadsheet** checkbox in Figure 8–23 is nonactive (grayed-out). Again, the reason for it being nonactive is that the **Print Preview** image is directly tied to the active on-screen image. To make this checkbox functional, the on-screen image in Figure 8–24 is changed for the **Gantt Chart** to include the **Activity Usage Spreadsheet** (bottom of the screen). Note that with this change, the **Spreadsheet** checkbox in Figure 8–25 is now visible.

Mark the **Resource Spreadsheet** checkbox (Figure 8–25) to include the profile in the **Print Preview**. Compare Figure 8–25, with the spreadsheet, to Figure 8–9, without the spreadsheet.

The last of the choices available under the **Options** tab of the **Page Setup** dialog box (Figure 8–17) is the **Page Settings** field. You can use the **Break Page Every Group** checkbox to start a new page, in the **Print Preview**, each time the group changes.

Default Button.   The **Page Setup** dialog box (Figure 8–17) has a number of button options available at the bottom of the dialog box to manage

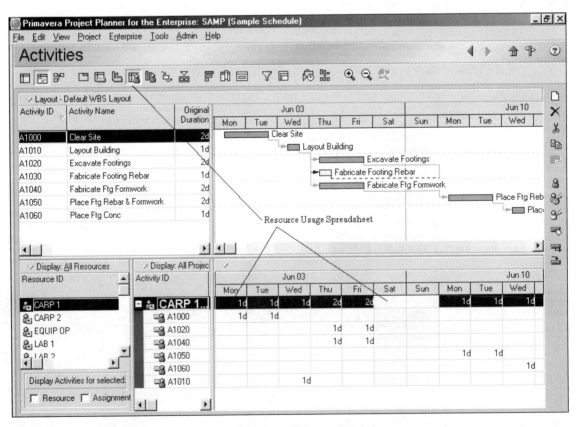

Figure 8–24   Gantt Chart Screen—**Resource Usage Spreadsheet**—Resources

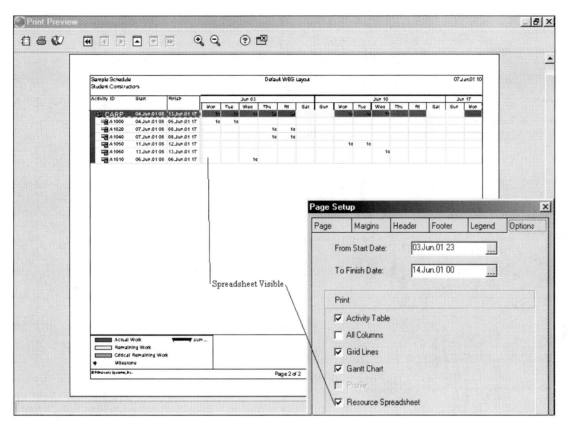

Figure 8–25   Options Tab—**Resource Spreadsheet**

the use of the settings provided in the dialog box. By clicking the **Default** button, the settings of the **Page Setup** dialog box will revert back to the default values provided by *P3e*'s Portfolio Analyst. All changes made to the **Page Setup** dialog box will be lost when the settings revert to the defaults.

Apply Button.   Click the **Apply** button to see your selections applied to the **Print Preview** without closing the **Page Setup** dialog box.

EXAMPLE PROBLEM

Figure 8–26 shows the Gantt chart hard-copy prints for a house put together as an example for student use (see the wood-framed house drawings in the Appendix).

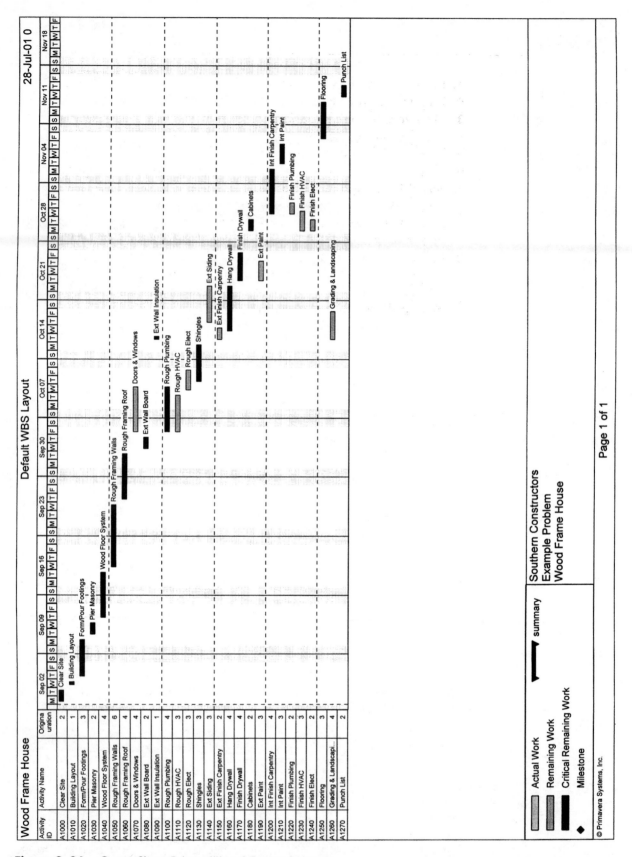

Figure 8–26 Gantt Chart Print—Wood-Framed House

EXERCISES

1. Prepare Gantt chart prints for Figure 3–12 (Exercise 3, Chapter 3).
2. Prepare Gantt chart prints for Figure 3–13 (Exercise 4, Chapter 3).
3. Prepare Gantt chart prints for Figure 3–14 (Exercise 5, Chapter 3).
4. Prepare Gantt chart prints for Figure 3–15 (Exercise 6, Chapter 3).
5. Prepare Gantt chart prints for Figure 3–16 (Exercise 7, Chapter 3).
6. Prepare Gantt chart prints for Figure 3–17 (Exercise 8, Chapter 3).
7. **Small Commercial Concrete Block Building—Gantt Chart Print**
   Prepare Gantt chart printouts for the small commercial concrete block building located in the Appendix. Prepare a hard-copy print using the **File** main pull-down menu option.
8. **Large Commercial Building—Gantt Chart Print**
   Prepare Gantt chart printouts for the large commercial building located in the Appendix. Prepare a hard-copy print using the **File** main pull-down menu option.

PERT Diagrams

Objectives

Upon completion of this chapter, you should be able to:

- Modify an on-screen PERT diagram
- Modify a PERT hard-copy print

VIEW THE PERT DIAGRAM

The PERT diagram (activity-on-node logic diagram) is a visual representation of the activities and shows the relationships among the activities. To see the PERT view of the sample schedule, click on the **PERT** button from the Activity Toolbar (Figure 9–1). The on-screen view changes from the Gantt chart format to the PERT format (Figure 9–1).

The arrows in this view show the relationships (predecessor/successor) among the activities. Activity A1000, Clear Site, is a predecessor to Activity A1010, Layout Building; therefore, Clear Site must be completed before Layout Building can begin. Activity A1010, Layout Building, is a successor to Activity A1000, Clear Site. Showing the relationships among activities on the screen makes them easier to understand. Each of the activities or rectangular shapes in Figure 9–1 is called a node and contains information about the activity.

Activity Details

You can edit an activity in the **PERT** view within *P3e* the same way as in the Gantt chart format. The easiest way is to access **Activity Details**, shown at the bottom of Figure 9–2. This form is accessed by clicking on the **Activity Details** button on the Activity Toobar (Figure 9–2).

With **Activity Details** selected, click on the activity to be edited. Figure 9–3 shows Activity A1000, Clear Site, with a bold box around it, meaning that it has been selected. The information appearing in **Activity Details** relates to the selected activity. Any of the **Activity Details** tabs and fields can be edited by simply clicking on the field and modifying the information.

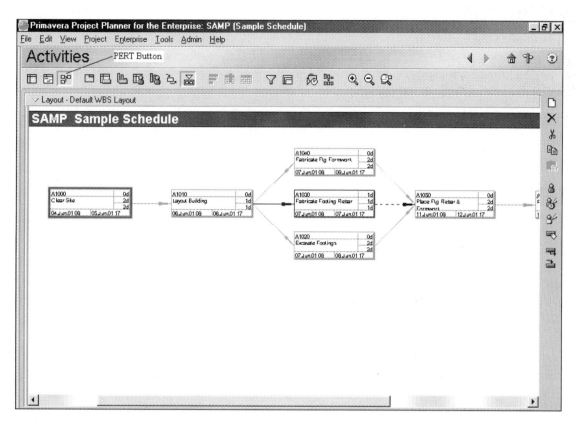

Figure 9–1   PERT View

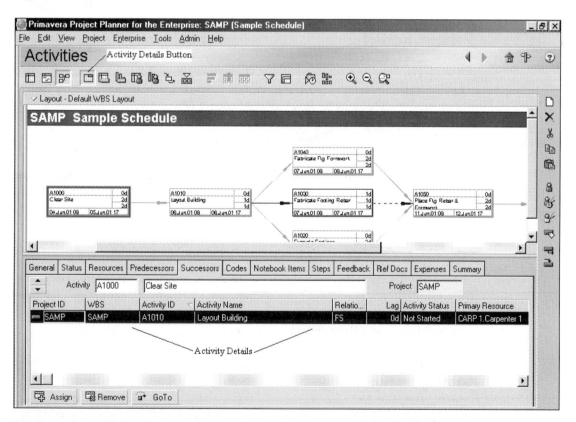

Figure 9–2   Activity Details

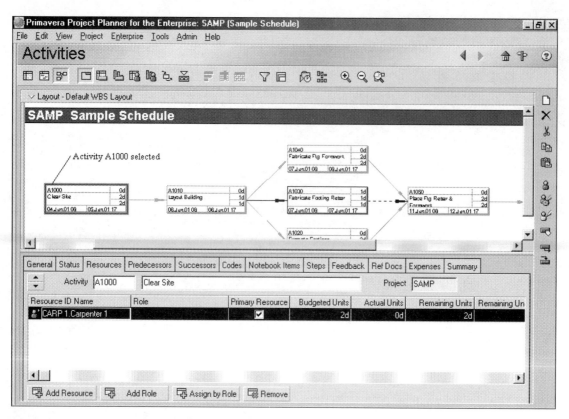

Figure 9–3 Activity Details—Activity A1000 Selected

Insert Activity

There are advantages to adding a new activity using the PERT view rather than the Gantt chart view. The primary advantage is in actually seeing the placement of the activity and its relationship to the other activities. To insert an activity, click on the PERT screen (Figure 9–4). Next, select the **Add** button from the **Edit** main pull-down menu.

In Figure 9–4, the new activity was inserted. Note in Figure 9–4 that the **Activity Details** is still active at the bottom on the screen. In Figure 9–5, the **Activity** field is input as A1045. *P3e* automatically assigns the number A1070, but that number doesn't relate to placement in our diagram. Since we set the interval at 10 for each new activity, there is room between established activities to insert a new activity. The new activity is set at 1045. Input the activity description (Order Concrete) field and the original **Duration** (1 day) next. When you click on any other activity button, the information from **Activity Details** is entered in the cells of the new activity box (Figure 9–5).

Move Activity

Many times the activity placement in the view automatically created by *P3e*, or the view that results when an activity is inserted, needs to be

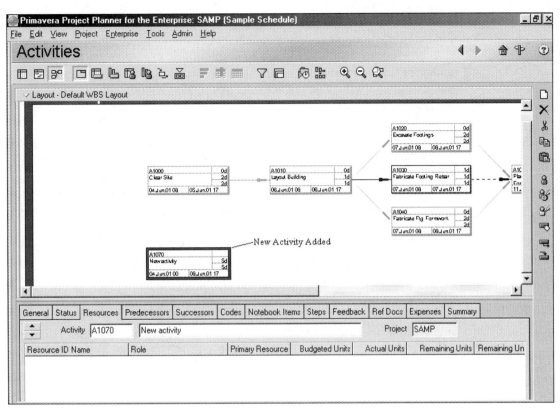

Figure 9–4   New Activity Added

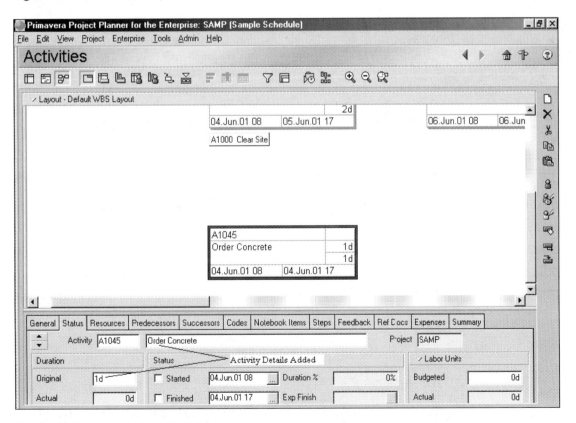

Figure 9–5   Activity Details Added

rearranged to be more meaningful. Moving activities to rearrange the *P3e* PERT view is very simple. Click on the activity to be moved and drag it to the new location. In Figure 9–4, the newly created Activity A1045 is selected. It has a dark border around it. Hold down the left mouse button and drag the box to the new location. Figure 9–6 shows the location to where the box was dragged.

The only relationship, so far, between the inserted activity and the rest of the diagram is the sequencing of the activity number. The relationships (predecessor/successor activities) between the inserted activity and the other activity need to be established. Activity A1010, Layout Building, needs to be established as a predecessor activity. Activity A1060, Place Ftg. Conc., needs to be established as a successor activity. To establish a SF (start–finish) relationship between Activity A1010 and the new activity (A1045), place the mouse pointer at the box end of Activity A1010 (Figure 9–7). The mouse pointer changes from a mouse arrow to a rleationship arrow to indicate that a relationship is being established. Hold the mouse button down, drag it to the beginning of Activity A1045, and release the mouse button. The new relationship line between the two activities has been established (Figure 9–7). Following the same pro-

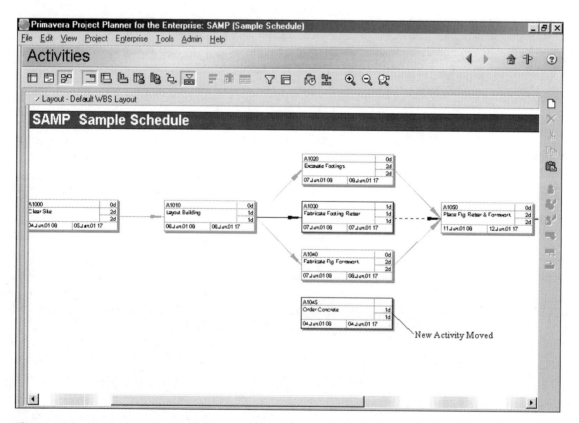

Figure 9–6   New Activity Moved

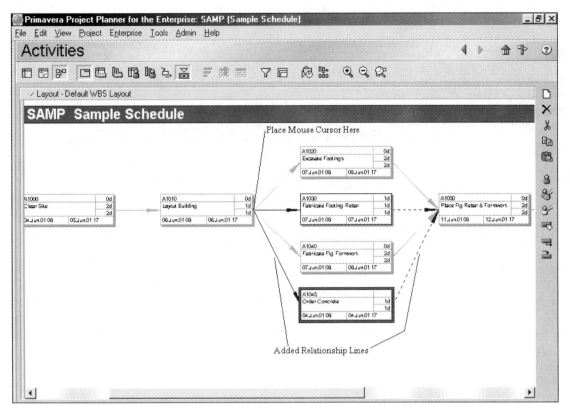

Figure 9–7   Relationship Lines Added

cedure, establish the FS relationship between Activities A1045 and A1060. Now the relationships between the inserted activity and the rest of the diagram have been established. If the added relationship needs to be edited, click on the relationship line to be edited to select it. Then after selecting the relationship, double click the left mouse button on the relationship line, and the **Edit Relationship** dialog box will appear (Figure 9–8). Using the **Edit Relationship** dialog box, the **Relationship Type** can be changed or **Lag Days** can be added.

Trace Logic

The **Trace Logic** feature, accessed from the **Trace Logic** button from the Activity Toolbar is a valuable tool when there are logic problems with a diagram. The Trace Logic view is not available with the Gantt chart format. Click on the activity you want (Activity A1030 in Figure 9–9) and **Trace Logic** displays all predecessors and successors to the selected activity. On the **Trace Logic** screen in the lower half of Figure 9–9, Activity A1030 has a box around it, designating it as the selected activity. Activity A1010 is shown as a predecessor activity and Activity A1050 is shown as a successor activity.

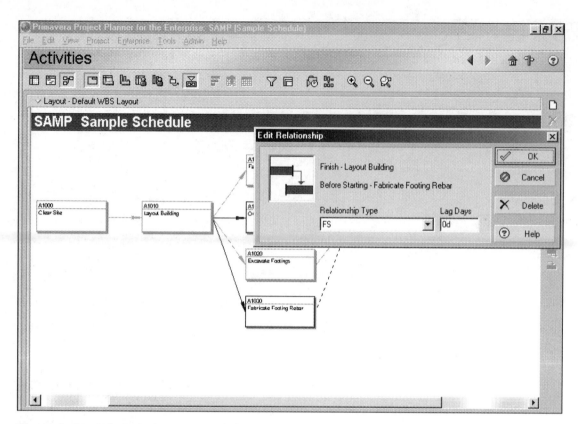

Figure 9–8   Edit Relationship Dialog Box

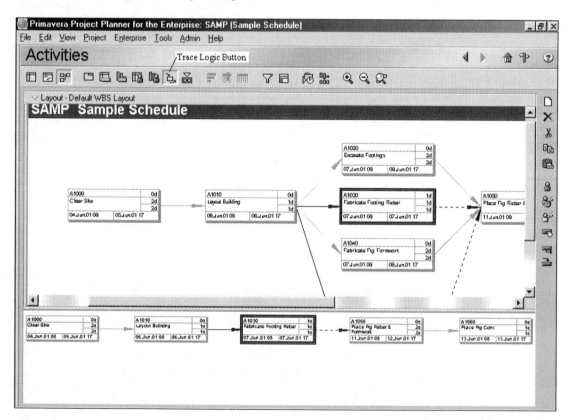

Figure 9–9   Trace Logic

Zoom

The zoom functions (**Zoom In**, **Zoom Out**, and **Zoom to Fit Best**) on the PERT view are the same as on the on-screen Gantt chart. The PERT diagram shown in Figure 9–10 is obtained by clicking on the **Zoom to Fit Best** button. Compare Figure 9–10 with Figure 9–1. Obviously, being able to place many activities on the same screen while still being able to read all pertinent information is a very valuable tool.

FORMAT THE PERT DIAGRAM

Many times for clarity, you may want to change the information in the node for the on-screen and the hard-copy print of the PERT diagram. *P3e* enables you to change the node using **PERT Options**.

Activity Box Configuration

To change the configuration of the node, select **PERT Options** from the **View** main pull-down menu (Figure 9–11). The **PERT Options** dialog

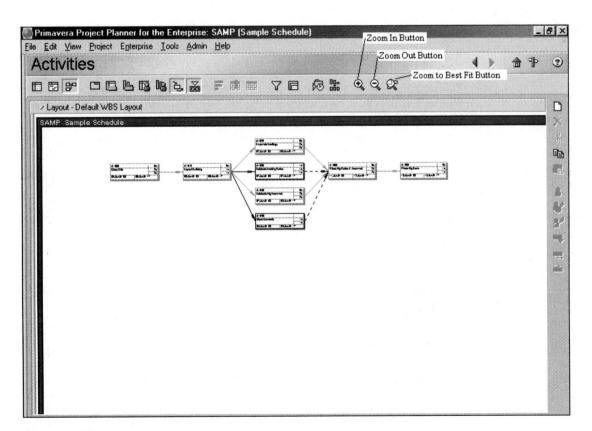

Figure 9–10   Zoom Buttons

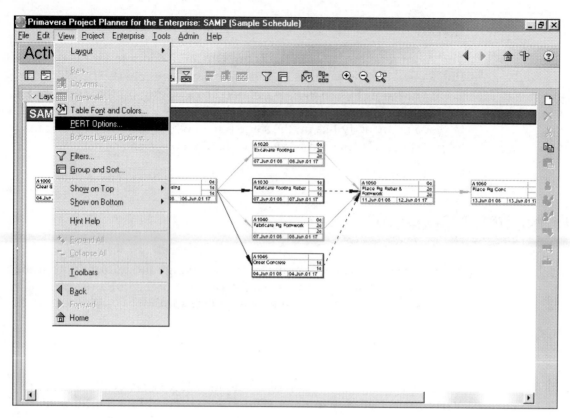

Figure 9–11   View Main Pull-Down Menu—**PERT Options**

box will appear (Figure 9–12). *Note: In the **Activity box templates** field there are many preconfigured templates for node format.* The **Durations** template is highlighted in Figure 9–12. This means that the nodes for the PERT view for the Sample Schedule are in this format. The preview field of the dialog box shows the information and configuration of the information contained in the **Activity box templates** field. Using multiple configurations, each for a particular need, is the primary purpose of this dialog box. The preconfigured templates are:

> **Activity Name**
> **Activity ID**
> **Durations**
> **Planned Dates**
> **Budgeted Units**
> **Budgeted Cost**
> **Resource View**
> **Current Status**

Figure 9–13 shows the **Activity Box Template** changed to the **Activity Name** template. Compare the **Activity Name** template that has been

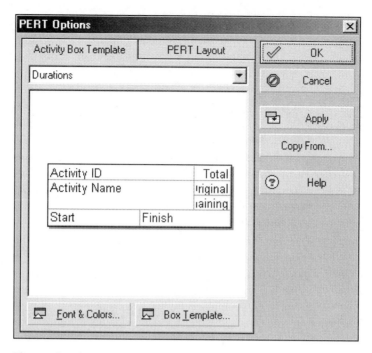

Figure 9–12   PERT Options Dialog Box

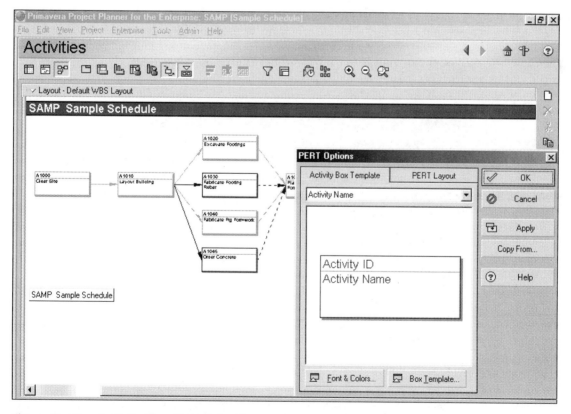

Figure 9–13   Activity Box Template Changed

applied to the PERT view in Figure 9–13 to the **Durations** template in Figure 9–11. Obviously, by limiting the information that appears in the activity box, many more activities can be placed in the PERT view.

The **PERT Options** dialog box also allows you to modify preconfigured templates to contain the fields and information you desire. Click on the **Box Template** button (Figure 9–14) to produce the **Chart Box Template** dialog box (Figure 9–15). Note from Figure 9–14 that the **Chart Box Template** dialog box has two fields (**Activity ID** and **Activity Name**) shown in the **Field Name** field. Three changes have been made to the **Chart Box Template** dialog box shown in Figure 9–15, from that shown in Figure 9–14, to demonstrate the use of this dialog box in changing activity box configurations.

The first change is that the locations of the **Activity ID** and **Activity Name** fields have been changed in Figure 9–15. To accomplish this, the **Activity ID** field is selected and then the **Shift down** button was clicked to shift its location with that of the **Activity Name** field. The **Shift up** and **Shift down** buttons are used to change the vertical location of fields with respect to each other.

The second change is the addition of a new field. To accomplish this, **Add** button is clicked to create the new blank field. Then, by clicking the mouse in the new blank field, a pop-up window appears for selecting the new field. **Early Start** is selected as the new field in Figure 9–15.

The third change is that the **Total Width** field has been changed from 130 to 110.

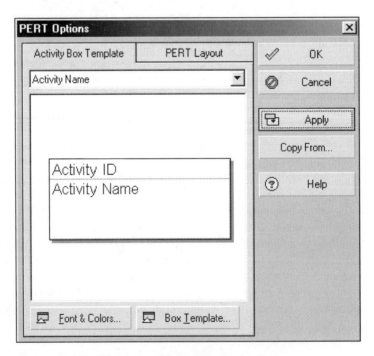

Figure 9–14   Chart Box Template Dialog Box

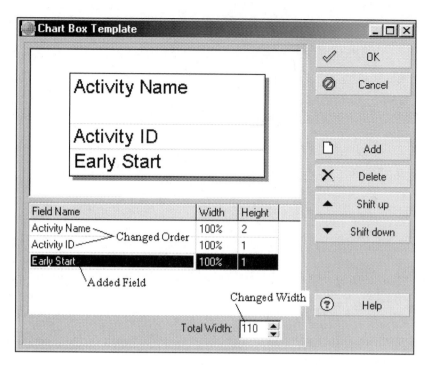

Figure 9–15   Chart Box Template Dialog Box—Changes Made

One last change made to the **Chart Box Template** dialog box is its reconfiguration. In Figure 9–16, the **Width** of both the **Activity ID** and the **Early Start** fields is changed to 50%. Note that with this width change, these fields are now placed side by side, horizontally.

Now click the **OK** button of the **Chart Box Template** dialog box to accept the changes. Compare the activity box configuration in the **Chart Box Template** dialog box **PERT Options** dialog box in Figure 9–17, with the activity box changes, to Figure 9–13, without the changes.

Fonts and Colors

Again, for clarity, you may want to change the activity box font (see Figure 9–1) and/or colors for the on-screen and the hard-copy print of the PERT diagram. To change either the font or colors of the activity box configuration of the PERT screen, select the **Font & Colors** button of the **PERT Options** dialog box (Figure 9–17). The **Chart Font and Colors** dialog box appears (Figure 9–18).

Font Button.   The first option available with the **Chart Font and Colors** dialog box is the **Font** button. Clicking the **Font** button produces the **Font** dialog box (Figure 9–19). Choosing the **Font**, **Font style**, **Size**, **Effects**, font **Color**, and **Script** are options available with this dialog box. Compare the **Font** dialog box shown in Figure 9–20 to that of

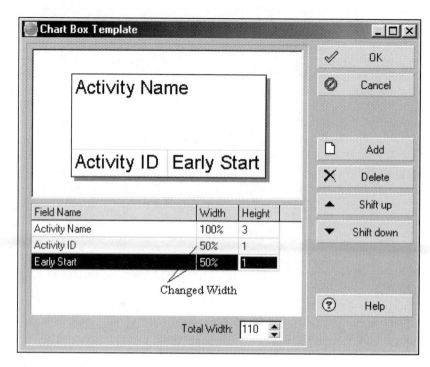

Figure 9–16   Chart Box Template Dialog Box—Changed Width

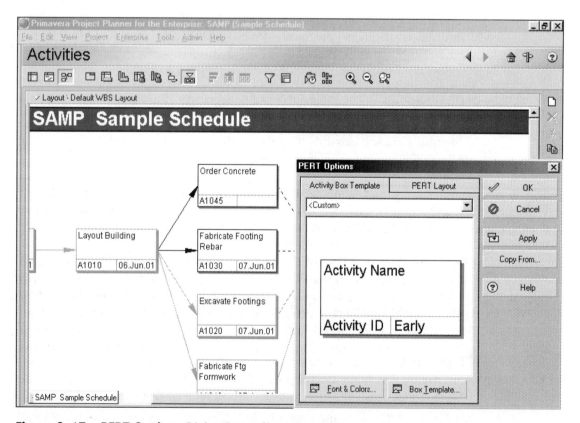

Figure 9–17   PERT Options Dialog Box—Changes Made

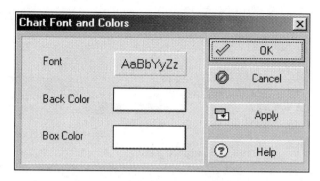

Figure 9–18   Chart Font and Colors Dialog Box

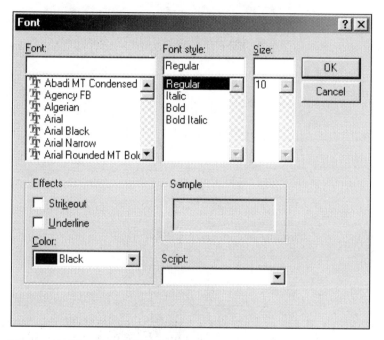

Figure 9–19   Font Dialog Box

Figure 9–19. The **Font**, **Font style**, and the **Underline Effects** fields are changed. Note that the **Sample** field provides a preview of the changes. Click the **OK** button to accept these changes.

Color Buttons.   The other options available with the **Chart Font and Colors** dialog box are the **Back Color** and the **Box Color** buttons (Figure 9–18). Clicking either of these buttons produces the **Color** dialog box shown in Figure 9–21 for changing color selection. The **Back Color** button is used to change the background color of the PERT screen and hardcopy prints. The **Box Color** button is used to change the activity box color.

Compare the activity box configuration of the **PERT Options** dialog box in Figure 9–22, with the font and color changes, to Figure 9–17, without the changes.

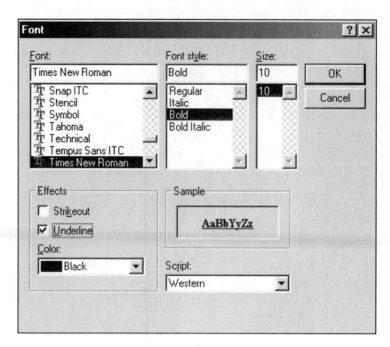

Figure 9–20 **Font** Dialog Box Changed

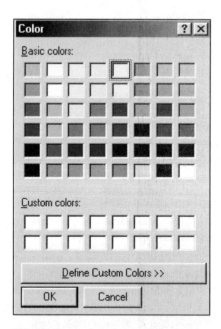

Figure 9–21 Color Dialog Box

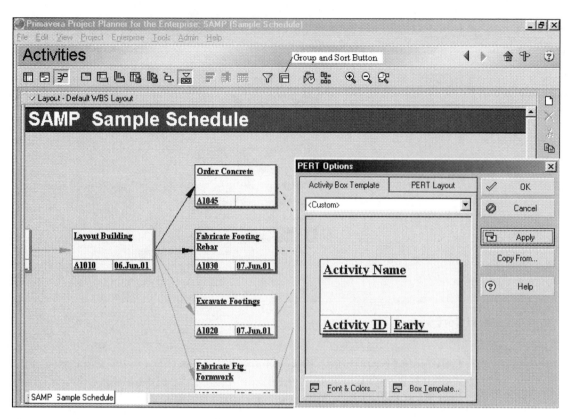

Figure 9–22   PERT Options Dialog Box—Changes Made

GROUP AND SORT

You may sometimes want to change the organization of the activities for the on-screen and the hard-copy print of the PERT diagram. You may want to group similar activities together, sorted by some set of criteria. For communication purposes, you may want to group all the activities of a particular subcontractor, crew, location, or responsibility together. To change the screen activity organization, select the **Group and Sort** button, from the Activity Toolbar (see Figure 9–22). The **Group and Sort** dialog box will appear (Figure 9–23). The grouping is defined in the **Group By** pull-down menu. Figure 9–22 was executed with **<None>** in the **Group By** field. In Chapter 5, **Activity Codes** (Figure 5–23) were defined so that the activities of the Sample Schedule could be grouped by responsibility. The **Activity Code** of **Responsibility** was created with **CARP** (Carpentry Crew) and **LAB** (Laborer Crew) as the responsibility grouping options. **Responsibility** is selected in the **Group By** field in the

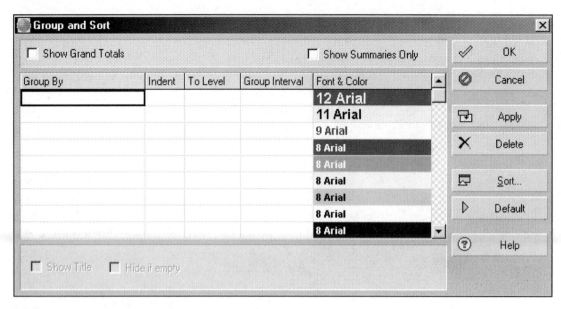

Figure 9–23   Group and Sort Dialog Box

Group and Sort dialog box in Figure 9–24 as the level 1 sort criteria. To accept the new grouping criteria, click the **Apply** button of the **Group and Sort** dialog box and the new groupings appears in the on-screen PERT view (Figure 9–24).

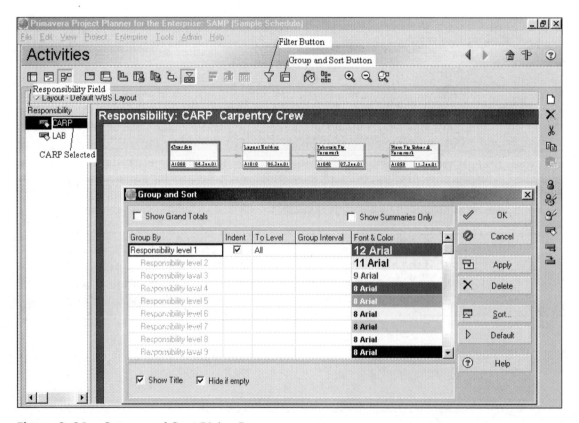

Figure 9–24   Group and Sort Dialog Box

Compare the PERT view that is grouped by responsibility (Figure 9–24) with the ungrouped PERT view (Figure 9–1). Note that the group title, **Responsibility: CARP Carpentry Crew**, appears at the top of the on-screen PERT view. The title can be hidden using the **Group and Sort** dialog box. Also note in the on-screen PERT view that a **Responsibility** field appears at the left of the on-screen view. Here the **Responsibility** field includes the grouping options of **CARP** and **LAB** (showing the hierarchy of the grouping).

In Figure 9–24, the **CARP** group is selected for display. To change to the **LAB** group, simply click on **LAB**. To remove the grouping criteria and have the on-screen PERT view appear as in Figure 9–1, select **<None>** in the **Group By** field of the **Group and Sort** dialog box. Then click the **Apply** button.

Filter

You may occasionally want to filter or limit the activities that appear on the on-screen and the hard-copy print of the PERT diagram. You may want to show only the activities filtered by some criterion. To filter the screen activity selection, select the **Filter** button (see Figure 9–24). The **Filters** dialog box will appear (Figure 9–25). The preconfigured filter options are shown in Figure 9–25.

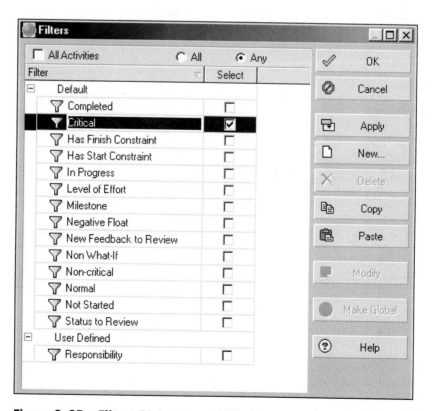

Figure 9–25   Filters Dialog Box—Critical Activities

Figure 9–25 shows **Critical** activities selected. Click on the **Apply** button. Figure 9–26 is the result. Since all the activities in the Sample Schedule are critical (0 total float) except Activity A1030, only Activity A1030 was "filtered out."

In Chapter 5 the **Filters** dialog box was used to create a **User Defined** filter criterion called **Responsibility** (Figure 5–31). The **Activity Code of Responsibility** was created with **CARP** (Carpentry Crew) and **LAB** (Laborer Crew) as the responsibility filtering options. **Responsibility** is selected in the **User Defined** filters (Figure 9–27) to filter activities for the Sample Schedule.

Selecting the **Modify** button of the **Filters** dialog box produces the **Filter** dialog box. In Figure 9–28, the **Responsibility** filter is modified to select only activities that have been identified as CARP activities in the **Activity Code** field. To accept the new filter criteria, click the **OK** button of the **Filter** dialog box and then the **Apply** button of the **Filters** dialog box; the new filter activities will then appear in the on-screen PERT view (Figure 9–29).

Compare the PERT view that is filtered by responsibility (Figure 9–29) with the unfiltered PERT view (Figure 9–1) and the filtered Gantt chart (Figure 5–32). In Figure 9–29, the **CARP** group is selected for display. To

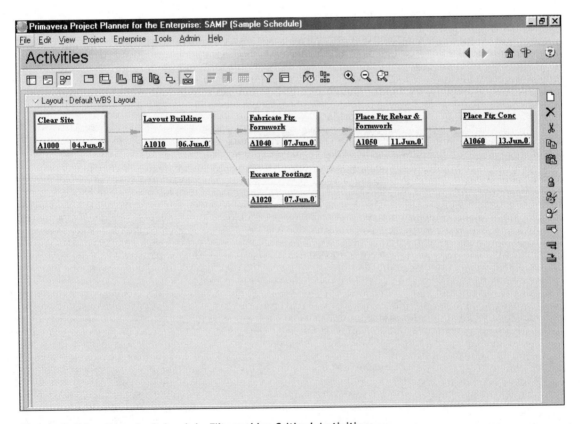

Figure 9–26   Sample Schedule Filtered by Critical Activities

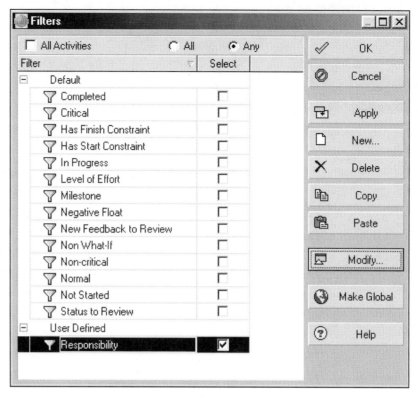

Figure 9–27   Filters Dialog Box

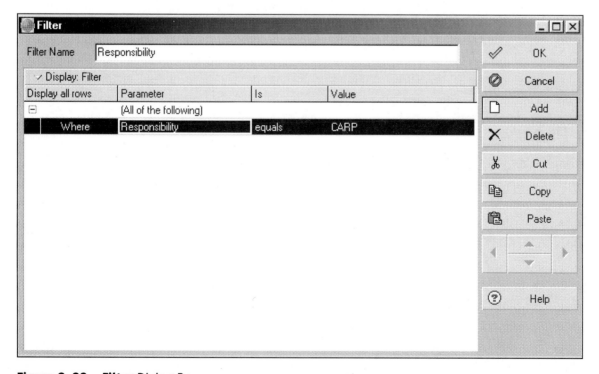

Figure 9–28   Filter Dialog Box

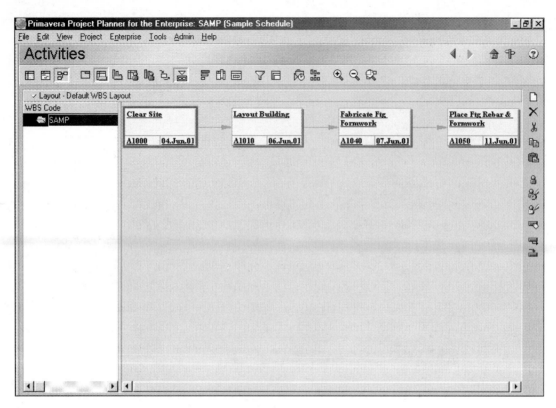

Figure 9–29   Sample Schedule Filtered by Responsibility

change to the **LAB** group, call the **Filter** dialog box back up and change the filter criteria. To remove the filtering criteria and have the on-screen PERT view appear as in Figure 9–1, uncheck all the checkboxes in the **Select** field of the **Filters** dialog box (Figure 9–27). Then click the **Apply** button.

PRINT PREVIEW

The print options for the PERT view are very similar to the Gantt chart view discussed in Chapter 8. The **Print Preview** for the PERT view is accessed from the **File** main pull-down menu. Figure 9–30 is the **Print Preview** screen for the Sample Schedule. Compare the PERT **Print Preview** screen (Figure 9–30) to the Gantt chart view (see Figure 8–2).

Toolbar

Note that all the Toolbar options for the two **Print Preview** screens are the same and have the same functions. Again, when the on-screen PERT

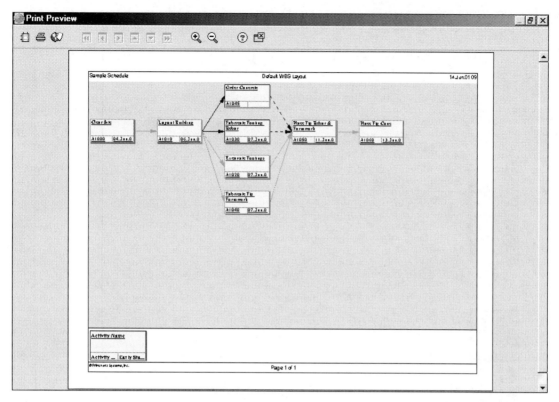

Figure 9–30   PERT—Print Preview

view is changed, the same change is carried over to the **Print preview** screen and to the hard-copy print. The **Print preview** options for the PERT view operate essentially in the same way as discussed in detail in Chapter 8.

Print

Figure 9–31 is a hard-copy print of the PERT view of the Sample Schedule.

Example Problem

Figure 9–32 shows two PERT hard-copy prints for a house put together as an example for student use (see the wood-framed house drawings in the Appendix).

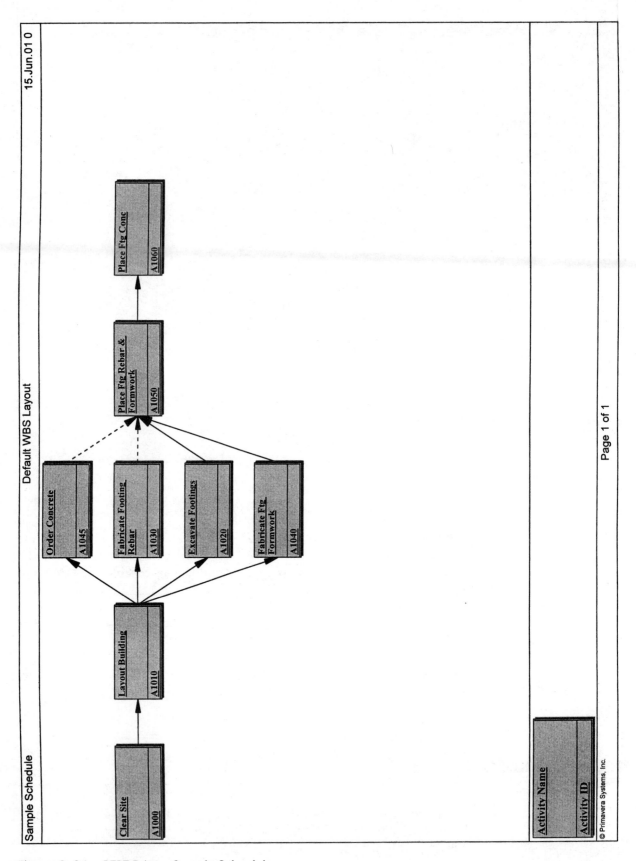

Figure 9–31 PERT Print—Sample Schedule

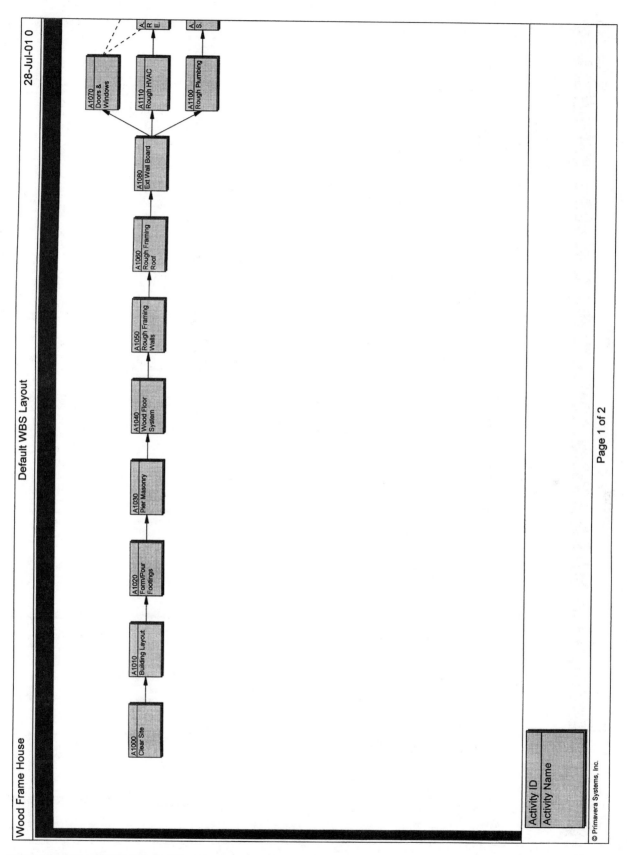

Figure 9–32a   PERT Print—Wood-Framed House

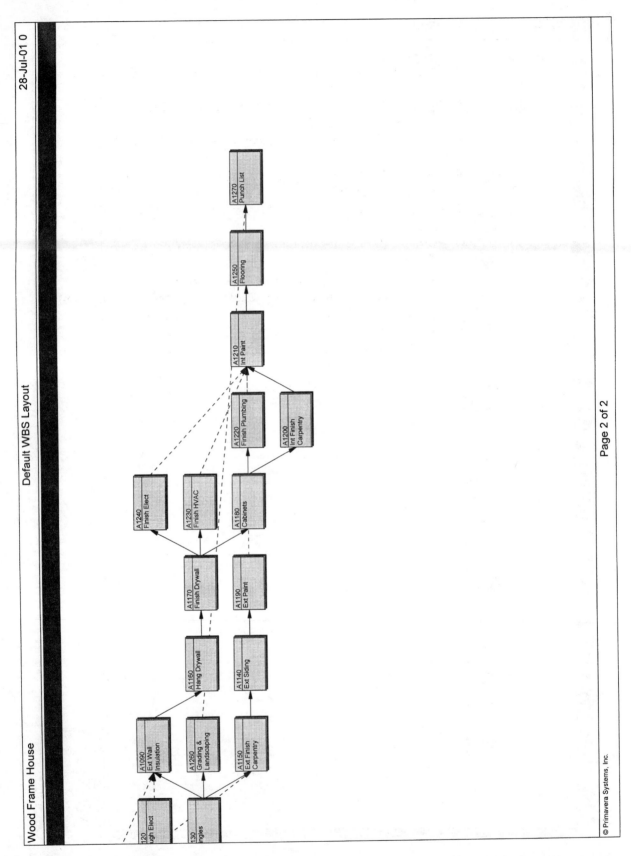

Figure 9–32b   PERT Print—Wood-Framed House

EXERCISES

1. Prepare PERT diagram hard-copy prints for Figure 3–12 (Exercise 3, Chapter 3).
2. Prepare PERT diagram hard-copy prints for Figure 3–13 (Exercise 4, Chapter 3).
3. Prepare PERT diagram hard-copy prints for Figure 3–14 (Exercise 5, Chapter 3).
4. Prepare PERT diagram hard-copy prints for Figure 3–15 (Exercise 6, Chapter 3).
5. Prepare PERT diagram hard-copy prints for Figure 3–16 (Exercise 7, Chapter 3).
6. Prepare PERT diagram hard-copy prints for Figure 3–17 (Exercise 8, Chapter 3).
7. **Small Commercial Concrete Block Building**
   Prepare PERT diagram hard-copy prints for the small commercial concrete block building located in the Appendix.
8. **Large Commercial Building**
   Prepare PERT diagram hard-copy prints for the large commercial building located in the Appendix.

10 Tabular Reports

Objectives

Upon completion of this chapter, you should be able to:

• Modify and print Tabular Reports

Sometimes a tabular report (table) is a more convenient format for transferring *P3e* update data. This chapter illustrates the presentation of the hard-copy tabular report in schedule, resource, and cost formats.

VALUE OF TABULAR REPORTS

Although there are many styles of *P3e* formatted tabular reports, a logic tabular report is chosen as a demonstration example. Tabular reports are very useful as a communication tool.

The tabular reports are accessed through **Reports** from the **Tools** main pull-down menu (Figure 10–1). The **Reports** screen will appear (Figure 10–2). The **Reports** screen lists all reports defined for this project (**SAMP**), as well as those that come with *P3e*. You can either delete or modify the specifications to existing reports, or add a new report title and specification.

Within the **Reports** screen, click on **LG-01 Logic Report By Project** to select it. Then, click the **Run** button from the **Reports** screen button bar (Figure 10–2). The **Run Report** dialog box will appear (Figure 10–3). The **Run Report** dialog box has print options. They are:

• View the on-screen **Print Preview.**
• Send the print file **Directly to Printer.**
• Send the print file to a directory. You can choose an **HTML File** or an **ASCII Text File** structure. Use the **Output** file button to tell *P3e* where to store the print file.

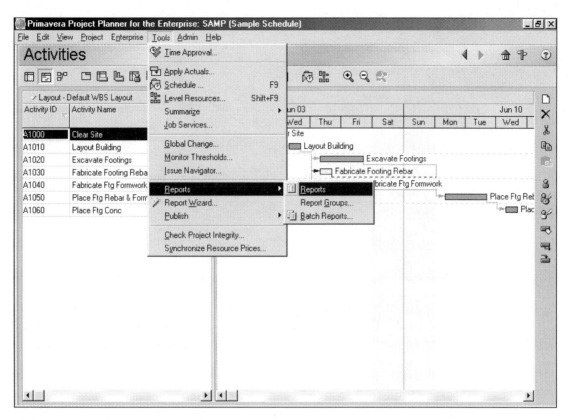

Figure 10–1   Tools Main Pull-Down Menu

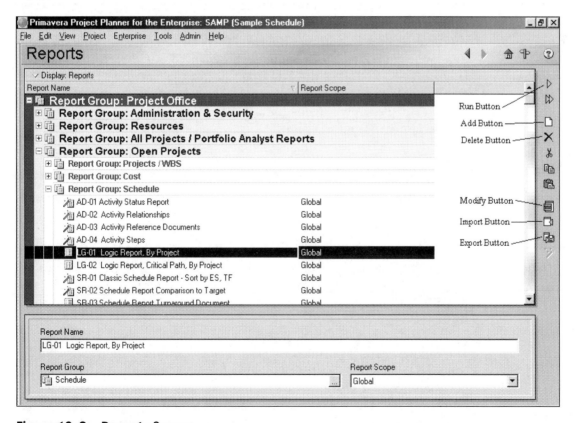

Figure 10–2   Reports Screen

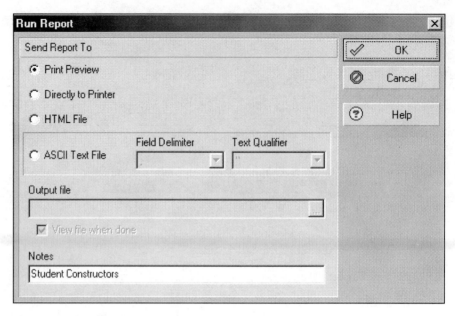

Figure 10–3   Run Report Dialog Box

In the **Run Report** dialog box in Figure 10–3, **Student Constructors** is entered into the **Notes** field.

In Figure 10–3, **Print Preview** is selected and the **OK** button is clicked. The **Print Preview** in Figure 10–4 is the result. Note the location of the **Student Constructors** information that was input in the **Notes** field of the **Run Report** dialog box (Figure 10–3). Compare the **Print Preview** available from the **Tools** main pull-down menu for tabular reports (Figure 10–4) to that of the **File** main pull-down menu for Gantt charts (Figure 8–2) and for PERT diagrams (Figure 9-30). Note that the format of the **Print Preview** screens for all three different report types are identical in appearance and usage.

Also observe from Figure 10–4 that the **Page Setup** button is available for changing header, footer, and print margin specifications. The use of the **Page Setup** dialog box will be discussed in greater detail later in this chapter.

Figure 10–5 is a copy of the actual hard-copy print of the **LG-01 Logic Report By Project**. This report is obtained by selecting the **Print** button from the **Print Preview** screen (Figure 10–4). This report is a hard-copy print of selected information from the *P3e* activity table data in tabular form.

The tabular form is particularly useful as a communication tool when it is necessary to give a craftsperson or subcontractor only the information pertinent to their work. A sort by the activities pertinent to their work, with only necessary information about relevant activities, will reduce confusion and should increase efficiency. Tabular reports can be used in

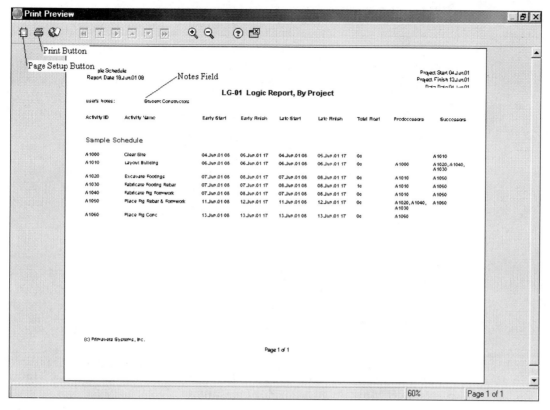

Figure 10–4   Print Preview Screen

Activity ID	Activity Name	Early Start	Early Finish	Late Start	Late Finish	Total Float	Predecessors	Successors
Sample Schedule								
A1000	Clear Site	04.Jun.01 08	05.Jun.01 17	04.Jun.01 08	05.Jun.01 17	0d		A1010
A1010	Layout Building	06.Jun.01 08	06.Jun.01 17	06.Jun.01 08	06.Jun.01 17	0d	A1000	A1020, A1040, A1030
A1020	Excavate Footings	07.Jun.01 08	08.Jun.01 17	07.Jun.01 08	08.Jun.01 17	0d	A1010	A1050
A1030	Fabricate Footing Rebar	07.Jun.01 08	07.Jun.01 17	08.Jun.01 08	08.Jun.01 17	1d	A1010	A1050
A1040	Fabricate Ftg Formwork	07.Jun.01 08	08.Jun.01 17	07.Jun.01 08	08.Jun.01 17	0d	A1010	A1050
A1050	Place Ftg Rebar & Formwork	11.Jun.01 08	12.Jun.01 17	11.Jun.01 08	12.Jun.01 17	0d	A1020, A1040, A1030	A1060
A1060	Place Ftg Conc	13.Jun.01 08	13.Jun.01 17	13.Jun.01 08	13.Jun.01 17	0d	A1050	

Sample Schedule
Report Date 19.Jun.01 09

Project Start 04.Jun.0
Project Finish 13.Jun.0
Data Date 04.Jun.0

LG-01 Logic Report, By Project

User's Notes: Student Constructors

Figure 10–5   Hard-Copy Print

conjunction with diagrams or used as stand-alone reports. They are also useful as worksheets when updating the schedule.

Note that the activities in Figure 10–5 are sorted by day (Early Start) and can have the same columns as appear on-screen if the columns are uncovered. Column selection and formatting were discussed in Chapter 5.

REPORT EDITOR

In addition to providing a large number of preconfigured reports (Figure 10–2), *P3e* provides two report addition/modification options that can be used to access and format information stored in the *P3e* database. These two report addition/modification options are the **Report Editor** and the **Report Wizard**.

The **Report Editor** is used to modify an existing *P3e*'s preconfigured report. It provides the capability to tailor a report to your specific requirements. It also enables you to group, sort, filter, and roll-up project information.

To use the **Report Editor**, select the report to be modified. Then click the **Modify** button from the **Reports** screen (Figure 10–2). The **Report Editor** screen for the selected report will appear. The title bar of the **Report Editor** screen displays the name of the currently selected report (**LG_01 Logic Report, By Project** in Figure 10–6).

The **Report Editor** screen includes five major report modification tools for you to use in editing reports. They are **Toolbar, Right-Click Menu, Ruler, Left Margin**, and the **Report Canvas**.

The first of the five **Report Editor** tools available for report modification comprises the **Toolbar** buttons (Figure 10–6). The **Toolbar** buttons include:

- **New Report Button.**   Click this button to create a new report. You are prompted to verify that you want to delete current report components.
- **Add Date Source Button.**   Click this button to create a new data source. A data source is a category of information. The **Detail Area** of the **Report Canvas** area is the only area of the report to which you can add a data source. Examples of data sources are activities, cost codes, expense categories, and work breakdown structure.
- **Add Row Button.**   Click this button to add a new row of information to the report. When you select the **Add Row** button, the new row is added to the report below the selected row. The **Properties** dialog box is immediately opened to define the properties of the new row. The **Properties** dialog box is discussed later in this chapter.
- **Add Text Cell Button.**   When you select the **Add Text Cell** button, a new text cell is added to the selected row. The **Properties** dialog box is immediately opened to define the properties of the new text cell.
- **Add Image Cell Button.**   When you select the **Add Image Cell** button, the new image cell is added to the selected row. The **Properties** dialog box is immediately opened to define the properties of the new image cell.
- **Add Line Cell Button.**   Click this button to add a new line cell to the report. The **Properties** dialog box is immediately opened to define the properties of the new line cell.

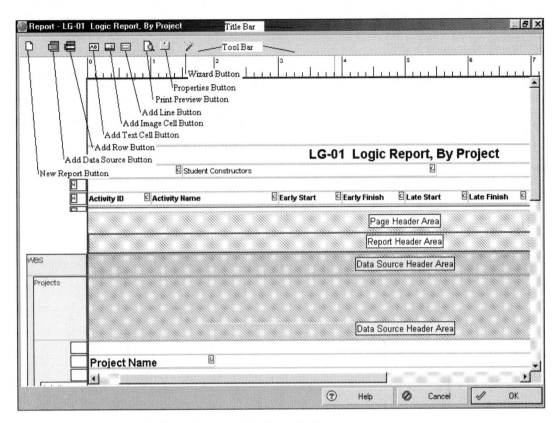

Figure 10–6   Report Editor Screen—**Toolbar** Buttons

- **Print Preview Button.**   Click this button to view the **Print Preview** before the report changes are accepted. When closing the **Report Editor** screen you will be asked if you want to save the changes made to the report. Being able to use the print preview to evaluate the changes before the report configuration modifications are saved is an extremely important editing tool.
- **Properties Button.**   Clicking this button produces the **Properties** dialog box for the selected field. The **Properties** dialog box is discussed later in this chapter.
- **Wizard Button.**   Clicking this button produces the **Report Wizard**. Use the **Report Wizard** to create new customized reports that are not found among *P3e*'s preconfigured reports (Figure 10–2). The **Report Wizard** is discussed later in this chapter.

The second of the five **Report Editor** tools available for report modification comprises the **Right-Click Menu** options (Figure 10–7). Select the cell to be modified. In Figure 10–7, the **User Notes** cell is selected. Then click the right mouse button to bring up a menu of field modification options. These modification options include: **Cut, Copy, Paste, Delete, Auto Arrange**, and access to the **Properties** dialog box.

The third of the five **Report Editor** tools available for report modification is the **Ruler** tool (Figure 10–7). The **Ruler** indicates the horizontal position of each report cell. A blue, shaded area indicates the position and width

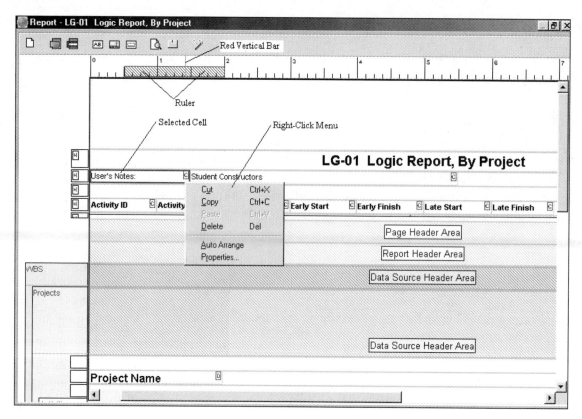

Figure 10–7   **Report Editor** Screen—**Right-Click Menu** and **Ruler** Options

of the selected cell. Note that, in Figure 10–7 the **Ruler** indicates the position and width of the **User Notes** cell. A red, vertical bar within the **Ruler** indicates the position of your cursor within the **Report Editor** screen.

The fourth of the five **Report Editor** tools available for report modification is the **Left Margin** tool (Figure 10–8). The **Left Margin** helps you identify the data source and row type. To identify data sources, the **Left Margin** displays each data source's name in the upper-left corner of **Left Margin** field. Again, a data source is a category of information. To identify row types, the **Left Margin** displays each row's type in the upper-left corner of the row. Note that the **H** icon in Figure 10–8 indicates header rows. An **F** icon indicates a footer row.

The fifth of the five **Report Editor** tools available for report modification is the **Report Canvas** (Figure 10–9). Use the **Report Canvas** to add report components, to specify the location of the added components, and to edit existing components. Note that the **Report Canvas** does not display the actual size that each report component will occupy in **Print Preview** and hard-copy prints of the edited report.

The **Report Canvas** has eight different areas for the input of report components: **Standard Page Header Area, Page Header Area, Report Header Area, Data Source Header Area, Detail Area, Report Footer Area, Report Footer Area, Page Header Area,** and **Standard Page Footer Area.**

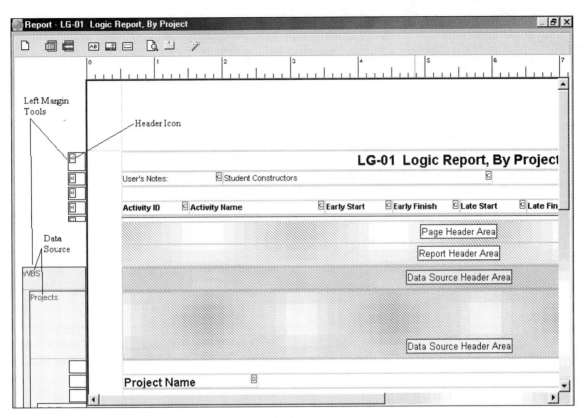

Figure 10–8   Report Editor Screen—**Left Margin** Tools

- **Standard Page Header/Footer Areas.**   Two of the eight **Report Canvas** areas are the **Standard Page Header/Footer Areas**. The only difference in how these two fields operate is that the **Standard Page Header Area** appears at the top of the compiled report and the **Standard Page Footer Area** appears at the bottom. The location of the **Standard Page Header Area** in the **Report Canvas** is shown in Figure 10–9. The location of the **Standard Page Header Area** in the **Print Preview** is shown in Figure 10–10. The **Standard Page Header/Footer Areas** are different from the **Page Header/Footer Areas**. The **Standard Page Header/Footer Areas** information comes from the **Page Setup** dialog box, which will be discussed later in this chapter. If you specify **Standard Page Header/Footer Area** information using the **Page Setup** dialog box, the **Standard Page Header/Footer** is displayed at the top and/or bottom of the compiled report, and will be followed by any information placed in the report's **Page Header/Footer Areas**. The advantage of using the **Standard Page Header/Footer Area** information over the **Page Header/Footer Areas** is that the same information is used for the other standard *P3e*'s preconfigured reports. Note that the information appearing in the **Standard Page Header Area** in Figure 10–9 resulted in the hard-copy print shown in Figure 10–5. No footer information was specified for the compiled report in Figure 10–5.
- **Page Header/Footer Areas.**   Two of the eight **Report Canvas** areas are the **Page Header/Footer Areas**. The location of the **Page Header**

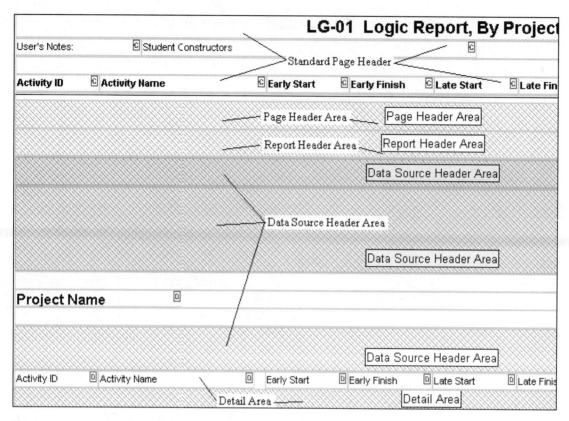

Figure 10–9   Report Canvas—Standard Page, Page, Report, and Data Source Headers Plus Detail Area Locations

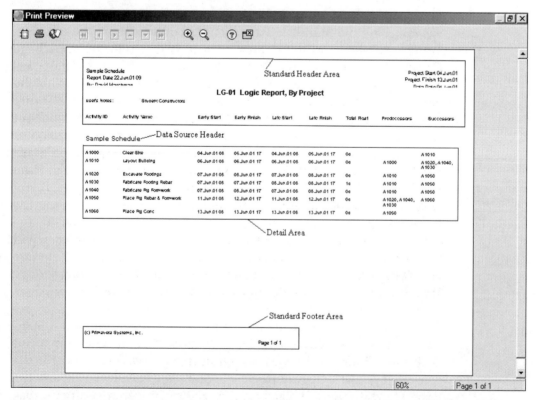

Figure 10–10   Print Preview—Standard Header, Footer, and Detail Area Locations

Area in the **Report Canvas** is shown in Figure 10–9. These two fields contain report information that appears at the top/bottom of each page of the compiled report. You can add rows and create variable and custom text cells within the added rows of the **Page Header/ Footer Areas.** Information appearing in the **Page Header/Footer Areas** applies only to the active report, and not to other standard *P3e* preconfigured reports.

*   **Report Header/Footer Areas.**   Two of the eight **Report Canvas** areas are the **Report Header/Footer Areas.** The location of the **Report Header Area** in the **Report Canvas** is shown in Figure 10–9. These two fields contain report information that appears at the top/bottom of only each first and last pages of the compiled report. You can add rows and create variable and custom text cells within the added rows of the **Report Header/Footer Areas**. Information appearing in the **Report Header/Footer Areas** applies only to the active report, and not to other standard *P3e* preconfigured reports.

*   **Data Source Header/Footer Areas.**   Two of the eight **Report Canvas** areas are the **Data Source Header/Footer Areas.** The location of the **Data Source Header Area** in the **Report Canvas** is shown in Figure 10–9. The location of the D**ata Source Header Area** in the **Print Preview** of the currently selected report (**LG_01 Logic Report, By Project**) is shown in Figure 10–10. These two fields contain report information that appears before/after the data source's records in the compiled report. You can add rows and create variable and custom text cells within the added rows of the **Data Source Header/Footer Areas.**

*   **Detail Area.**   The last of the eight **Report Canvas** areas is the **Detail Area**. The location of the **Data Source Header Area** in the **Report Canvas** is shown in Figure 10–9. The **Detail Area** contains the bulk of the project data shown on the compiled report. The location of the **Detail Area** in the **Print Preview** of the currently selected report (**LG_01 Logic Report, By Project**) is shown in Figure 10–10. The **Detail Area** of the **Report Canvas** area is the only area of the report to which you can add a data source. The data source indicates which category of the database information the report compiles. Examples of data sources are activities, cost codes, expense categories, and work breakdown structures. After selecting a data source, you add rows, and then cells within rows, to the **Detail Area**. The cells specify the individual data fields you want to report for the data source selected.

The **Report Canvas** provides visual cues to identify report component properties. The **Report Canvas** text cells are coded in a box in the upper-right corner of the field to indicate the type of data they report. There are four types of text cells: **Custom Text, Field Data, Filed Title,** and **Variable** cells.

*   **Custom Text Cell.**   A **C** in a box in the upper-right corner of the field indicates a **Custom Text** cell. A **Custom Text** cell contains text that you input for the cell. See Figure 10–11 for an example of a **Custom Text** cell. The text, **Student Constructors**, is input in a **Custom Text** cell.

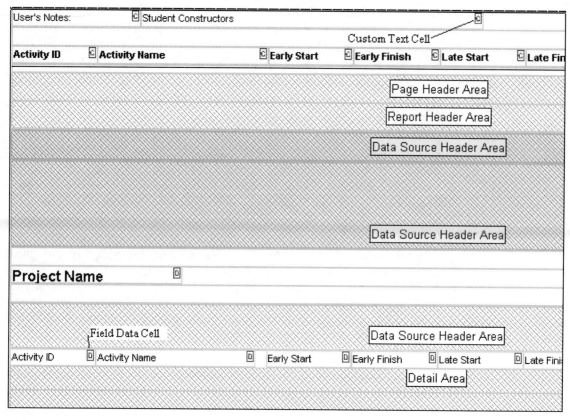

Figure 10–11   Report Canvas—Custom Text and Field Data Cells

- **Field Data Cell.**   A **D** in a box in the upper-right corner of the field indicates a **Field Data** cell. A **Field Data** cell contains information from the database that you specify. See Figure 10–11 for an example of a **Field Data** cell. The text, **Activity ID**, is input in a **Field Data** cell.
- **Field Title Cell.**   A **T** in a box in the upper-right corner of the field indicates a **Field Title** cell. A **Field Title** cell contains the name of a database field that you specify.
- **Variable Cell.**   A **V** in a box in the upper-right corner of the field indicates a **Variable** cell. A **Variable** cell contains information that relates to the overall report, rather than a specific data source, such as the report date or page number.

A **Report Canvas** text cell displayed in red indicates that the cell's properties either have not been defined or that there is a conflict in specifications.

PROPERTIES

Properties are characteristics of a report component, such as type and field for a text cell, sort and group options for a data source, and height for a row. To access the **Properties** dialog box, click the field of the **Report Canvas** to be modified. The **User's Notes**: custom text cell is selected in Figure 10–12. Note that a blue border surrounds the selected component. To increase the scope of the selected area, press the keyboard **Esc** key and the row will be selected.

With the field to be modified selected, the **Properties** dialog box is accessed by clicking the **Properties** button from the Button Bar of the **Report Editor** screen (Figures 10–6 and 10–12). The **Properties** dialog box can also be accessed by either double clicking in the selected field, or right clicking the selected field and then choosing **Properties**. The **Properties** dialog box is produced (Figure 10–13).

The **Properties** dialog box has six tabs containing **General Settings** options for modifying the fields of the compiled report. The six tabs of the **Properties** dialog box are **Report, Source, Row, Cell, Line,** and **Image**.

- **Cell** tab.   The first of the six **Properties** dialog box tabs is the **Cell** tab (Figure 10–13). The **User's Notes**: custom text cell is selected in Figure 10–12. The **Cell type** field (Figure 10–13) shows that **Custom Text** is selected for the field type. A **C** contained in a box in the upper-right corner of the field indicates a **Custom Text** cell (Figure 10–12). By clicking the down arrow, the **Cell type** can be changed to a **Field Data** cell (**D**), a **Field Title** cell (**T**), or a **Variable** cell (**V**) depending on the field location within the **Report Canvas**.

   The **General Settings** of the **Cell** tab gives you the ability to change the appearance of the cell information in the compiled report. You can change the cell custom text, cell alignment, text alignment, cell font, and borders. The **Custom Text** field of the **Cell** tab displays the text to appear in the selected **Report Canvas** field. Note that the **Custom Text** field is changed from **User's Notes**: in Figure 10–13 to **Group**: in Figure 10–14. Note also that the **Text alignment** field is changed from **Left Justify** in Figure 10–13 to **Center** in Figure 10–14.

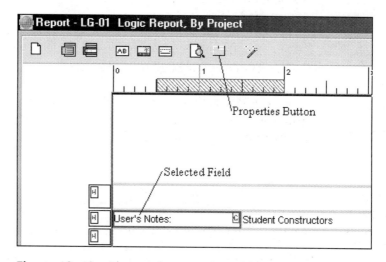

Figure 10–12   Report Canvas—Properties Button

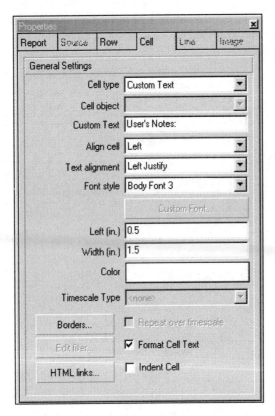

Figure 10–13 **Properties** Dialog Box—**Cell** Tab

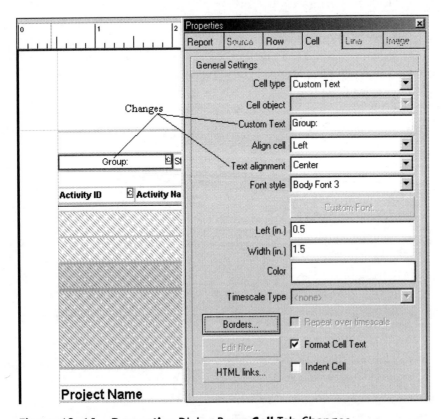

Figure 10–14 **Properties** Dialog Box—**Cell** Tab Changes

- **Row** tab. The second of the six **Properties** dialog box tabs is the **Row** tab (Figure 10–15). Note that the row **Height** and **Color** of the selected component can be changed using this tab.

- **Report** tab. The third of the six **Properties** dialog box tabs is the **Report** tab. In Figure 10–16, the **Title** field is blank. If, instead of being in the present field being modified (**Group:** field in Figure 10–14), we were in the report title field (Figure 10–8), then **LG-01 Logic Report** would appear in the **Title** field. Note that this is the same title that appears in the **Reports** (Figure 10–2), the **Print Preview** (Figure 10–4), and the **Report Editor** (Figure 10–6) screens for the selected report.

  The **General Settings** of the **Report** tab contains four buttons for modifying the selected report. The first of the **Report** tab buttons (Figure 10–16) is the **Background Image** button. This button is used to locate the file of an image to be used in the background of the hard-copy print of the selected report.

  The second of the **Report** tab buttons (Figure 10–16) is the **Report font styles** button. Clicking this button produces the **Edit font styles** dialog box (Figure 10–17). Note that in the **Cell** tab of the **Properties** dialog box (Figure 10–14), **Body Font 3** is the selected font style for the selected field. By selecting **Body Font 3** in the **Edit** font styles dialog box (Figure 10–17), the font style can be changed for the selected field (**Group:**).

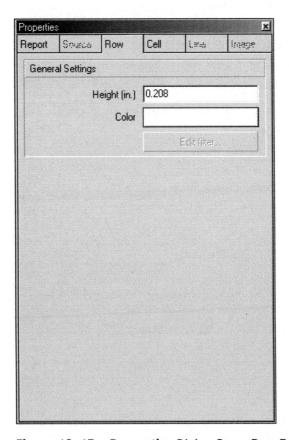

Figure 10–15   Properties Dialog Box—**Row** Tab

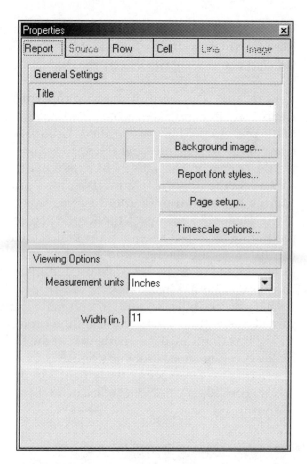

Figure 10–16 **Properties** Dialog Box—**Report** Tab

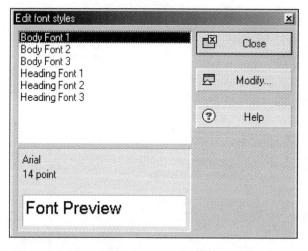

Figure 10–17 **Edit Font Styles** Dialog Box

The third of the **Report** tab buttons (Figure 10–16) of the **Properties** dialog box is the **Page setup** button. The **Page setup** button provides page format options for the hard-copy print and will be discussed in detail later in this chapter.

The fourth of the **Report** tab buttons (Figure 10–16) of the **Proper-ties** dialog box is the **Timescale options** button. Clicking this button produces the **Timescale options** dialog box (Figure 10–18). The pur-pose of the **Timescale options** dialog box is to allow you to modify and/or limit the timescale of the hard-copy print of the selected tab-ular report. As an example, if you wanted only to look at a single month's print of a 12-month schedule, you could use the **From Start Date** and the **To Finish Date** fields to limit the print length. These fields have nothing to do with the calculated dates within *P3e*. The dates input here only limit the length of the hard-copy print.

- **Line** tab. The fourth of the six **Properties** dialog box tabs is the **Line** tab (Figure 10–19). Note that the **Line** tab in Figure 10–14 is nonusable (grayed-out) for the selected field (**Group**:). To use this field, a new line is created. In Figure 10–19, a field is selected and the **Add Line Cell** button is clicked. The **Line** tab of the **Properties** dialog box immediately comes up to define the properties of the new line. The new line can be broken into multiple new cells.

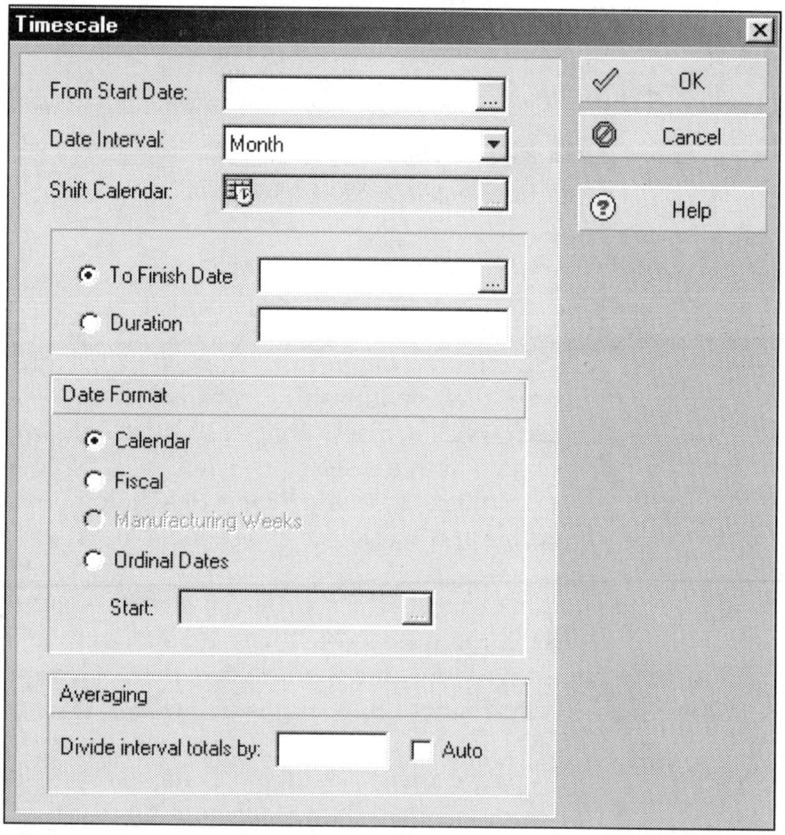

Figure 10–18 Timescale Dialog Box

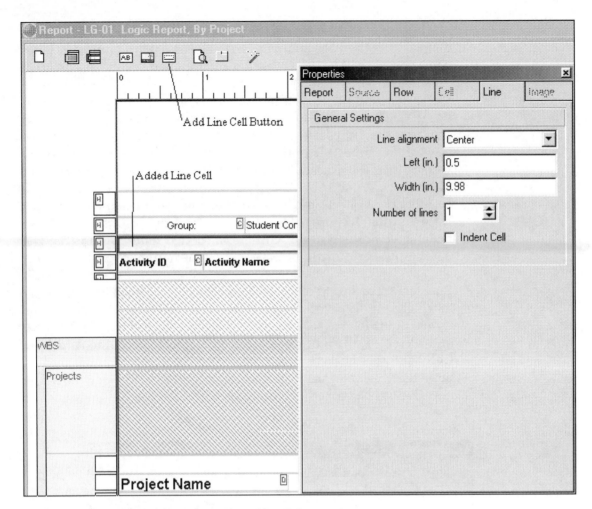

Figure 10–19 **Properties** Dialog Box—**Line** Tab

- **Image** tab. The fifth of the six **Properties** dialog box tabs to be shown is the **Image** tab (Figure 10–20). Note that the **Image** tab in Figure 10–14 is nonusable (grayed-out) for the selected field (**Group:**). To use this field, a new image cell is created. In Figure 10–20, the new line created above is selected. Then the **Add Image Cell** button is clicked. The **Image** tab of the **Properties** dialog box immediately comes up to define and locate the image file to be inserted in the new cell.

- **Source** tab. The sixth and last of the six **Properties** dialog box tabs is the **Source** tab. Note that the **Source** tab in Figure 10–14 is nonusable (grayed-out) for the selected field (**Group:**). The reason is that the selected field (**Group:**) is a **Custom Text** cell (**C**). The **Source Tab** is only available for **Field Data** cells (**D**) of the **Detail Area** field of the **Report Canvas**. The **Field Data** cell specifies the data you select for the report field. Note that in the **Activity ID** field of Figure 10–21, a **Field Data** cell (**D**) is selected. The **Activity ID Ascending** is infor-

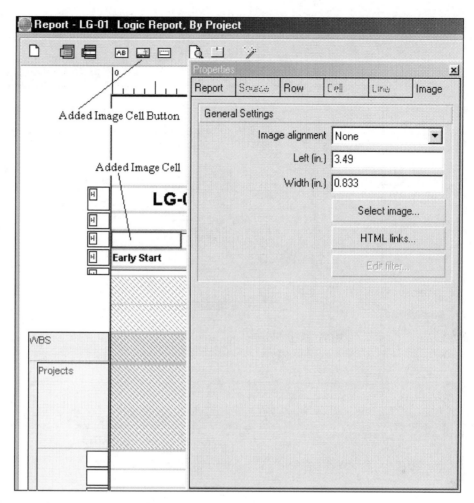

Figure 10–20   Properties Dialog Box—**Image** Tab

mation selected for presentation in the field. To specify other information for presentation in the selected field, click the **Modify** button, and the **Edit Sort** dialog box will appear (Figure 10–22). Use the **Sort by Object** and the **Sort by Field** fields to select the data to appear in the **Detail Area** field of the **Report Canvas**.

PAGE SETUP

The page setup function under the tabular reports also works the same way as the Gantt chart and PERT hard-copy prints. From the **Print Preview** screen (Figure 10–23) or the **Report** tab of the **Properties** dialog

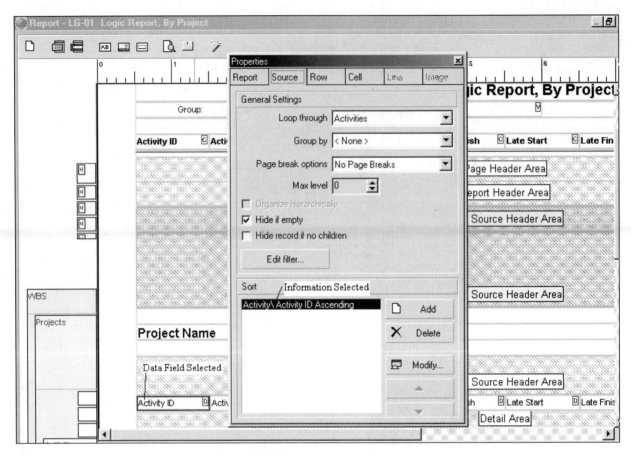

Figure 10–21   Properties Dialog Box—**Source** Tab

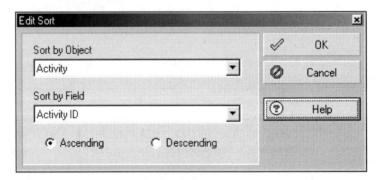

Figure 10–22   Edit Sort Dialog Box

box (Figure 10–16), select the **Page Setup** button, and the **Page Setup** dialog box will appear (Figure 10–23). The **Page Setup** dialog box has four tabs for report print modification: **Page, Margin, Header,** and **Footer.** The format and function of the four tabs of the **Page Setup** dialog box for *P3e*'s tabular reports is the same as for the Gantt chart and the PERT

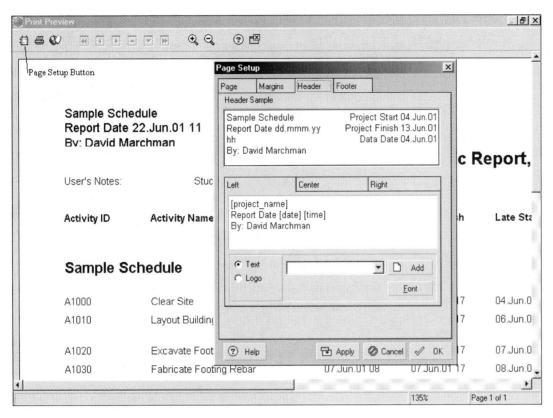

Figure 10–23   Page Setup Dialog Box—**Header** Tab

diagram. To view the **Page** tab of the **Page Setup** dialog box see Figure 8–6. To view the **Margins** tab see Figure 8-8. To view the Header tab see Figure 10–23 and Figure 8–11. To view the **Footer** tab of the **Page Setup** dialog box see Figure 8–12.

Changes Made to Tabular Report Using the Report Editor

Changes made to the selected report that was used in this chapter (**LG-01 Logic Report, By Project**) are shown in Figure 10–24. Compare Figure 10–24, the **Print Preview** made after the changes, to Figure 10–4, made before the changes. There are three changes made to the **Standard Header**. The first change is that a new line is added and the text **By: David Marchman** is added. The second change is that the **By Project** is removed from the **LG-01 Logic Report, By Project** title. The third change is that the text **User's Notes:** is changed to **Group:**. All three of these changes to the **Standard Header** were made using the **Custom Text** field of the **Cell** tab of the **Properties** dialog box.

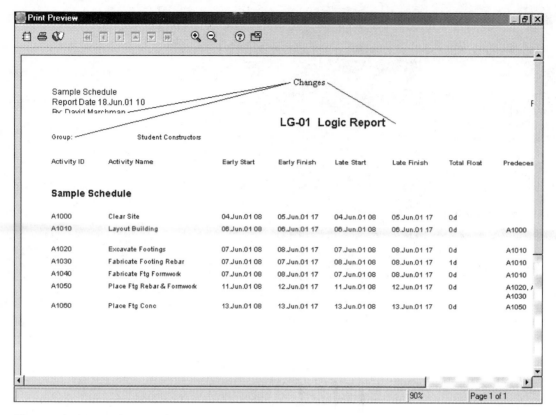

Figure 10–24 Print Preview—Changes Made to **Report Editor**

REPORT WIZARD

The **Report Editor** is a powerful editing tool, but *P3e*'s **Report Wizard** offers even more editing power. The **Report Wizard** is used to create new reports and modify existing wizard reports. The **Report Wizard** also enables you to:

- Select the specific data fields you want to include in the report
- Group, sort, and filter the report data
- Modify report column widths
- Add a report title
- Preview the report and modify it before saving it

To use the **Report Wizard**, select the report to be modified. The **SR-01 Classic Schedule Report – Sort by ES, TF** report is selected from the **Reports** screen (Figure 10–25). Figure 10–26 is a copy of the **Print Preview** run before any changes are made to the selected report. Click the **Run** button from the **Reports** screen (Figure 10–25) to obtain the **Print Preview**.

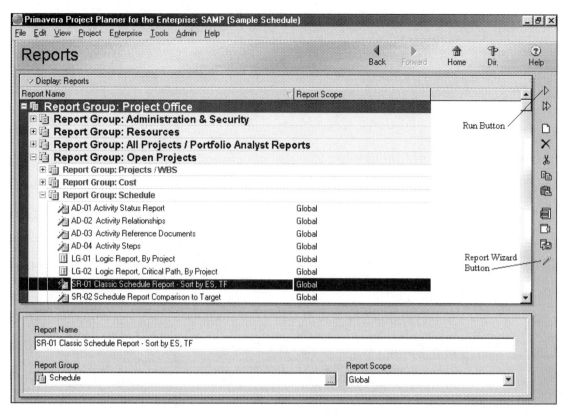

Figure 10–25 Reports Screen—**Report Wizard** Button

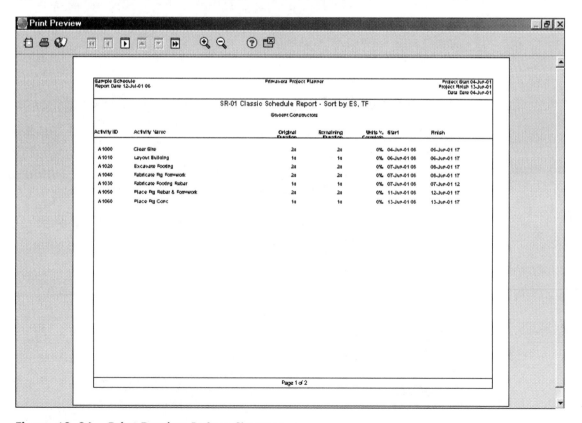

Figure 10–26 Print Preview Before Changes

Create or Modify Report.   Click the **Report Wizard** button from the **Reports** screen and the **Create or Modify Report** screen (Figure 10–27) of the **Report Wizard** is brought up. This screen is used to let *P3e* know whether you will be creating an entirely new report or modifying an existing report. Because an existing report was selected when the Report Wizard was accessed, only the **Modify Wizard Report** checkbox is active. Click the **Next** button to go to the **Select Subject Area** screen.

Select Subject Area.   The **Select Subject Area** screen (Figure 10–28) of the **Report Wizard** is used to select the portion of the report to be modified. The **Activities** selection is picked as the subject area for making report modifications. Click the **Next** button to go to the **Fields** screen.

Fields.   The **Fields** screen (Figure 10–29) of the **Report Wizard** is used to select the specific fields to be included in the report to be modified. Selecting the **Fields** button from the **Fields** screen produces the **Columns** dialog box (Figure 10–30) for selecting the columns to be included in the selected report. Note that the **Columns** dialog box for the **Report Wizard** is the same as used for formatting the Gantt view (Figure 5–6). The **Display Columns** field of the **Columns** dialog box shows the data columns that will appear in the modified report. Note that the columns appearing in the **Columns** dialog box (Figure 10–30) correspond to the selected report before modification (Figure 10–26). In Figure 10–31, the center left and right directional arrows are used to add/remove some of the displayed columns. The **Remaining Duration** column is removed and the **Budgeted Total Cost** column is added in its place. Click the **Next** button to go to the **Group & Sort** screen.

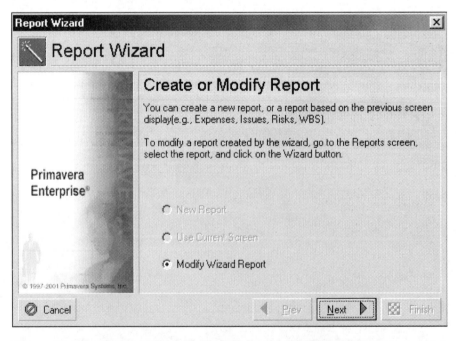

Figure 10–27   Report Wizard—Create or Modify Report Screen

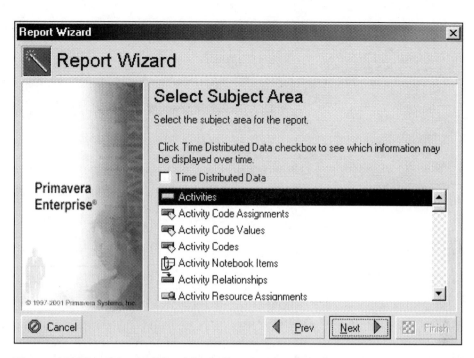

Figure 10–28   Report Wizard—Select Subject Area Screen

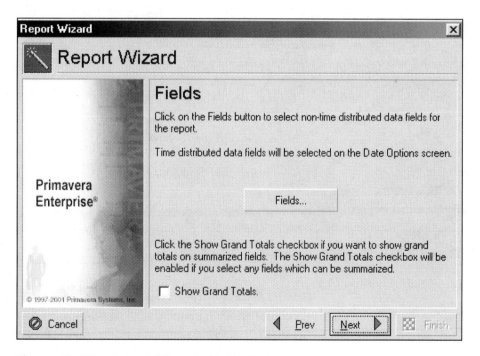

Figure 10–29   Report Wizard—Fields Screen

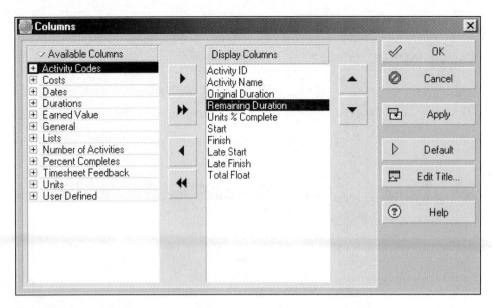

Figure 10–30   **Columns** Dialog Box

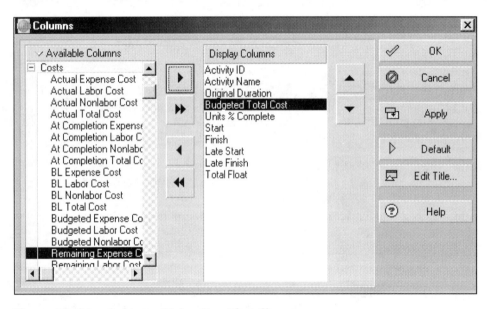

Figure 10–31   **Columns** Dialog Box After Changes

Group & Sort.   The **Group & Sort** screen (Figure 10–32) of the **Report Wizard** is used to group and sort the information included in the specific fields of the modified report. Selecting the **Group & Sort** button from the **Group & Sort** screen produces the **Group and Sort** dialog box (Figure 10–33) for grouping the information included in the modified report. Note that the **Group and Sort** dialog box for the **Report Wizard** is the same as that used for grouping/sorting the information in the **Gantt** view (Figure 5–25). The **Group By** field shows the grouping/sorting parameters appearing in the selected report. Note that the selected report has been changed to sort by **Responsibility** (Figure 10–33). Click the **Next** button to go to the **Filter** screen.

Figure 10–32   Report Wizard—Group & Sort Screen

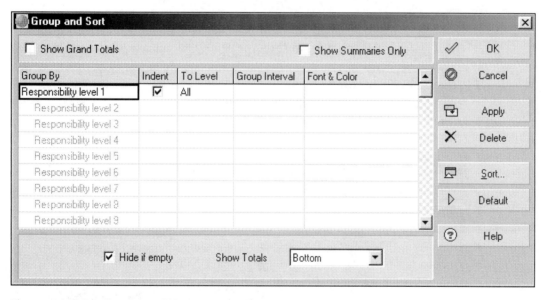

Figure 10–33   Group and Sort Dialog Box

Filter.   The **Filter** screen of the **Report Wizard** is used to limit the information included in the specific fields by some criteria. Selecting the **Filter** button from the **Filter** screen produces the **Filter** dialog box for selecting the limiting criteria for the modified report. Note that the **Filter** dialog box for the **Report Wizard** is the same as that used for filtering the information in the **Gantt** view (Figure 5–30). Click the **Next** button to go to the **Column Size** screen.

Column Size.   The **Column Size** screen (Figure 10–34) of the **Report Wizard** is used to modify the width of columns included in the modified report. Note in Figure 10–26 that the **Activity Name** column has wasted space. Select the **Activity Name** column from the **Column Size** screen (Figure 10–34) and drag it to resize it and eliminate the wasted space. Click the **Next** button to go to the **Report Title** screen.

Report Title.   The **Report Title** screen (Figure 10–35) of the **Report Wizard** is used to save the modified report under a name different from the original. Note in Figure 10–35 that the **Report Title** is changed from **SR-01 Classic Schedule Report – Sort by ES, TF** (Figure 10–25) to **SR-01 Classic Schedule Report – Modified** (Figure 10–35). Click the **Next** button to go to the **Report Generated** screen.

Report Generated.   The **Report Generated** screen of the **Report Wizard** is used to fine-tune the modified report before it. Use the **Run Report** button to produce a **Print Preview**. Figure 10–36 is a copy of the **Print Preview** including the changes made to the **SR-01 Classic Schedule Report – Sort by ES, TF**. Cancel the **Print Preview** to return to the **Report Generated** screen. Use the **Prev** and **Next** buttons of the **Report Generated** screen to return and modify any of the report selections until the report is formatted exactly the way you want it. Click the **Next** button to go to the **Congratulations** screen.

Congratulations.   The **Congratulations** screen is the last screen of the **Report Wizard** and is used to save the new modified report. Click the **Save Report** button to save the report under the new report title and return to the **Reports** screen (Figure 10–37). Note the location of the **SR-01 Classic Schedule Report – Modified** report.

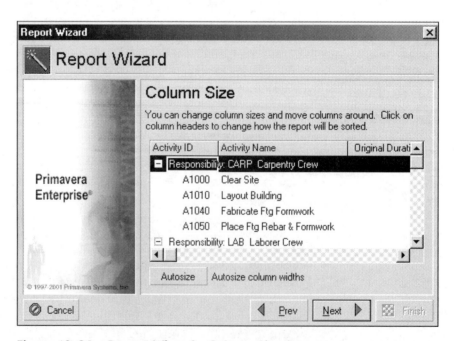

Figure 10–34   Report Wizard—Column Size Screen

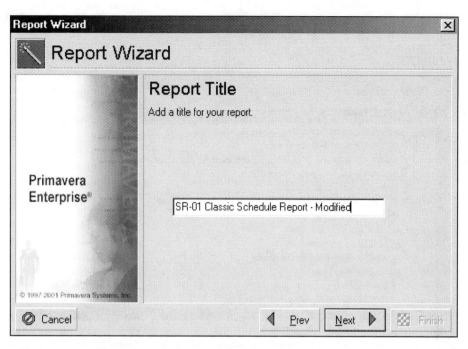

Figure 10–35   Report Wizard—Report Title Screen

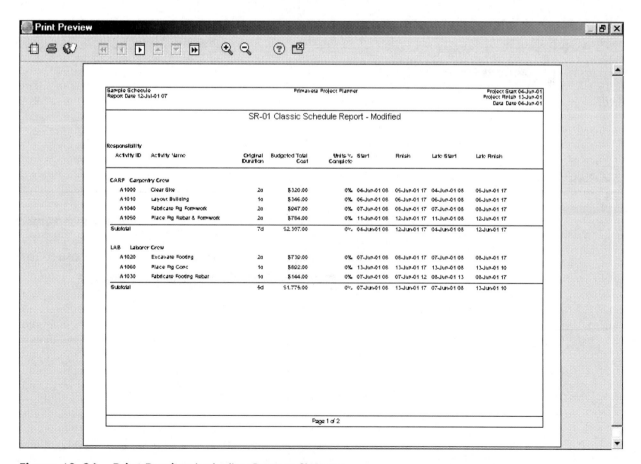

Figure 10–36   Print Preview Including Report Changes

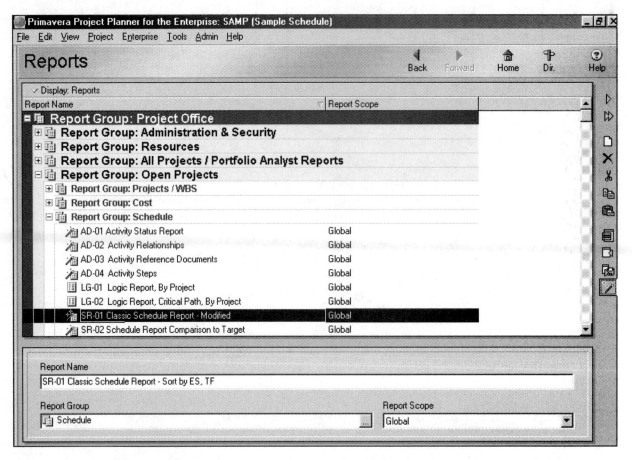

Figure 10–37   Reports Screen

EXAMPLE PROBLEM

Figures 10–38, 10–39, and 10–40 are tabular reports for the wood-framed house drawings in the Appendix. Figure 10–38 is a tabular report in the classic schedule format. Figure 10–39 is a tabular report in resource format. Figure 10–40 is a tabular report in cost format.

Wood Frame House
Report Date 28-Jul-01 07

Primavera Project Planner

Project Start 03-Sep-01
Project Finish 15-Nov-01
Data Date 03-Sep-01

SR-01 Classic Schedule Report - Modified

Activity ID	Activity Name	Original Duratio	Budgeted Total Cost	Units % Complete	Start	Finish	Late Start	Late Finish	Total Float
A1000	Clear Site	2	$1,280.0	0%	03-Sep-01 08	04-Sep-01 17	03-Sep-01 08	04-Sep-01 17	0
A1010	Building Layout	1	$386.0	0%	05-Sep-01 08	05-Sep-01 17	05-Sep-01 08	05-Sep-01 17	0
A1020	Form/Pour Footings	3	$1,174.0	0%	06-Sep-01 08	10-Sep-01 17	06-Sep-01 08	10-Sep-01 17	0
A1030	Pier Masonry	2	$967.0	0%	11-Sep-01 08	12-Sep-01 17	11-Sep-01 08	12-Sep-01 17	0
A1040	Wood Floor System	4	$4,181.0	0%	13-Sep-01 08	18-Sep-01 17	13-Sep-01 08	18-Sep-01 17	0
A1050	Rough Framing Walls	6	$3,323.0	0%	19-Sep-01 08	26-Sep-01 17	19-Sep-01 08	26-Sep-01 17	0
A1060	Rough Framing Roof	4	$3,468.0	0%	27-Sep-01 08	02-Oct-01 17	27-Sep-01 08	02-Oct-01 17	0
A1080	Ext Wall Board	2	$736.0	0%	03-Oct-01 08	04-Oct-01 17	03-Oct-01 08	04-Oct-01 17	0
A1090	Ext Wall Insulation	1	$385.0	0%	16-Oct-01 08	16-Oct-01 17	16-Oct-01 08	16-Oct-01 17	0
A1100	Rough Plumbing	4	$750.0	0%	05-Oct-01 08	10-Oct-01 17	05-Oct-01 08	10-Oct-01 17	0
A1130	Shingles	3	$1,091.0	0%	11-Oct-01 08	15-Oct-01 17	11-Oct-01 08	15-Oct-01 17	0
A1160	Hang Drywall	4	$1,844.0	0%	17-Oct-01 08	22-Oct-01 17	17-Oct-01 08	22-Oct-01 17	0
A1170	Finish Drywall	4	$790.0	0%	23-Oct-01 08	26-Oct-01 17	23-Oct-01 08	26-Oct-01 17	0
A1180	Cabinets	2	$1,618.0	0%	29-Oct-01 08	30-Oct-01 17	29-Oct-01 08	30-Oct-01 17	0
A1200	Int Finish Carpentry	4	$1,472.0	0%	31-Oct-01 08	05-Nov-01 17	31-Oct-01 08	05-Nov-01 17	0
A1210	Int Paint	3	$4,725.0	0%	06-Nov-01 08	08-Nov-01 17	06-Nov-01 08	08-Nov-01 17	0
A1250	Flooring	3	$1,583.0	0%	09-Nov-01 08	13-Nov-01 17	09-Nov-01 08	13-Nov-01 17	0
A1270	Punch List	2	$0.00	0%	14-Nov-01 08	15-Nov-01 17	14-Nov-01 08	15-Nov-01 17	0
A1110	Rough HVAC	3	$1,168.0	0%	05-Oct-01 08	09-Oct-01 17	08-Oct-01 08	10-Oct-01 17	1
A1120	Rough Elect	3	$940.0	0%	10-Oct-01 08	12-Oct-01 17	11-Oct-01 08	15-Oct-01 17	1
A1220	Finish Plumbing	2	$3,000.0	0%	31-Oct-01 08	01-Nov-01 17	02-Nov-01 08	05-Nov-01 17	2
A1070	Doors & Windows	4	$3,995.0	0%	05-Oct-01 08	10-Oct-01 17	10-Oct-01 08	15-Oct-01 17	3
A1230	Finish HVAC	3	$3,506.0	0%	29-Oct-01 08	31-Oct-01 17	01-Nov-01 08	05-Nov-01 17	3
A1240	Finish Elect	2	$1,410.0	0%	29-Oct-01 08	30-Oct-01 17	02-Nov-01 08	05-Nov-01 17	4
A1140	Ext Siding	3	$1,710.0	0%	18-Oct-01 08	22-Oct-01 17	29-Oct-01 08	31-Oct-01 17	7
A1150	Ext Finish Carpentry	2	$736.0	0%	16-Oct-01 08	17-Oct-01 17	25-Oct-01 08	26-Oct-01 17	7
A1190	Ext Paint	3	$525.0	0%	23-Oct-01 08	25-Oct-01 17	01-Nov-01 08	05-Nov-01 17	7
A1260	Grading & Landscaping	4	$600.0	0%	16-Oct-01 08	19-Oct-01 17	08-Nov-01 08	13-Nov-01 17	17d

Page 1 of 1

Figure 10–38 Tabular Report in Schedule Format for Wood-Framed House

Wood Frame House
Report Date 28-Jul-01 07

Project Start 03-Sep-01
Project Finish 15-Nov-01
Data Date 03-Sep-01

RA-01 Resource Assignments, All Activities

User's Notes: Student Constructors

Resource Project Code	Resource Name WBS Code	Activity Code	Activity	Planned Unit	Planned Duration	Planned Start	Planned Finish
CARPENTR Carpenter							
Example		A1010	Building Layout	2	1	05-Sep-01 08	05-Sep-01 17
Example		A1020	Form/Pour Footings	6	3	06-Sep-01 08	10-Sep-01 17
Example		A1040	Wood Floor System	8	4	13-Sep-01 08	18-Sep-01 17
Example		A1050	Rough Framing Walls	12d	6	19-Sep-01 08	26-Sep-01 17
Example		A1060	Rough Framing Roof	8	4	27-Sep-01 08	02-Oct-01 17
Example		A1070	Doors & Windows	8	4	05-Oct-01 08	10-Oct-01 17
Example		A1080	Ext Wall Board	4	2	03-Oct-01 08	04-Oct-01 17
Example		A1140	Ext Siding	6	3	18-Oct-01 08	22-Oct-01 17
Example		A1150	Ext Finish Carpentry	4	2	16-Oct-01 08	17-Oct-01 17
Example		A1200	Int Finish Carpentry	8	4	31-Oct-01 08	05-Nov-01 17
			Total	66			
CARP FOR Carpentry Forman							
Example		A1010	Building Layout	1	1	05-Sep-01 08	05-Sep-01 17
Example		A1020	Form/Pour Footings	3	3	06-Sep-01 08	10-Sep-01 17
Example		A1040	Wood Floor System	4	4	13-Sep-01 08	18-Sep-01 17
Example		A1050	Rough Framing Walls	6	6	19-Sep-01 08	26-Sep-01 17
Example		A1060	Rough Framing Roof	4	4	27-Sep-01 08	02-Oct-01 17
Example		A1070	Doors & Windows	4	4	05-Oct-01 08	10-Oct-01 17
Example		A1080	Ext Wall Board	2	2	03-Oct-01 08	04-Oct-01 17
Example		A1140	Ext Siding	3	3	18-Oct-01 08	22-Oct-01 17

Page 1 of 5

(c) Primavera Systems, Inc.

Figure 10–39 Tabular Report in Resource Format for Wood-Framed House

Wood Frame House
Report Date 28-Jul-01 07

AC-02 Activity Costs

Student Constructors

WBS

Activity ID	Activity Name	Activity Status	BL Total Cost	Actual Total Cost	Remaining Total Cost
Example					
	Wood Frame House				
A1000	Clear Site	Not Started	$1,280.00	$0.0	$1,280.00
A1010	Building Layout	Not Started	$386.00	$0.0	$386.00
A1020	Form/Pour Footings	Not Started	$1,174.00	$0.0	$1,174.00
A1030	Pier Masonry	Not Started	$967.00	$0.0	$967.00
A1040	Wood Floor System	Not Started	$4,181.00	$0.0	$4,181.00
A1050	Rough Framing Walls	Not Started	$3,323.00	$0.0	$3,323.00
A1060	Rough Framing Roof	Not Started	$3,468.00	$0.0	$3,468.00
A1070	Doors & Windows	Not Started	$3,995.00	$0.0	$3,995.00
A1080	Ext Wall Board	Not Started	$736.00	$0.0	$736.00
A1090	Ext Wall Insulation	Not Started	$385.00	$0.0	$385.00
A1100	Rough Plumbing	Not Started	$750.00	$0.0	$750.00
A1110	Rough HVAC	Not Started	$1,168.00	$0.0	$1,168.00
A1120	Rough Elect	Not Started	$940.00	$0.0	$940.00
A1130	Shingles	Not Started	$1,091.00	$0.0	$1,091.00
A1140	Ext Siding	Not Started	$1,710.00	$0.0	$1,710.00
A1150	Ext Finish Carpentry	Not Started	$736.00	$0.0	$736.00
A1160	Hang Drywall	Not Started	$1,844.00	$0.0	$1,844.00
A1170	Finish Drywall	Not Started	$790.00	$0.0	$790.00
A1180	Cabinets	Not Started	$1,618.00	$0.0	$1,618.00
A1190	Ext Paint	Not Started	$525.00	$0.0	$525.00
A1200	Int Finish Carpentry	Not Started	$1,472.00	$0.0	$1,472.00
A1210	Int Paint	Not Started	$4,725.00	$0.0	$4,725.00
A1220	Finish Plumbing	Not Started	$3,000.00	$0.0	$3,000.00

Page 1 of 4

Figure 10–40   Tabular Report in Cost Format for Wood-Framed House

EXERCISES

1. Prepare a classic schedule tabular hard-copy print for Figure 3–12 (Exercise 3, Chapter 3).
2. Prepare a classic schedule tabular hard-copy print for Figure 3–13 (Exercise 4, Chapter 3).
3. Prepare a classic schedule tabular hard-copy print for Figure 3–14 (Exercise 5, Chapter 3).
4. Prepare a classic schedule tabular hard-copy print for Figure 3–15 (Exercise 6, Chapter 3).
5. Prepare a classic schedule tabular hard-copy print for Figure 3–16 (Exercise 7, Chapter 3).
6. Prepare a classic schedule tabular hard-copy print for Figure 3–17 (Exercise 8, Chapter 3).
7. **Small Commercial Concrete Block Building—Tabular Reports**
   Prepare tabular reports for the small commercial concrete block building located in the Appendix.
8. **Large Commercial Building—Tabular Reports**
   Prepare tabular reports for the large commercial building located in the Appendix.

Section 3

Controlling

11 | Updating the Schedule

Objectives

Upon completion of this chapter, you should be able to:

- Copy a *P3e* schedule
- Establish a baseline schedule
- Record progress
- Apply actuals and establish a new data date
- Input duration changes
- Input logic changes
- Input added activities
- Produce updated reports

COMMUNICATING CURRENT INFORMATION

Schedule management can be summarized in the four following steps:

- Planning
- Scheduling
- Monitoring/updating (tracking and inputting schedule progress)
- Controlling (which is based on monitoring and suggests corrective action, if needed, to reflect the current situation)

Very few projects ever go exactly according to the original plan or schedule. Among the influences likely to change the original plan are weather, acts of God, better or worse productivity than anticipated, delivery problems, labor problems, changes in the scope of the project, interferences between craftpersons or subcontractors, and mismanagement in the flow of work. Once a project is under way, it is necessary to modify the schedule to keep it current.

When you update a project, you show that time has passed and progress has been attained on the project. *P3e* calculates the effects of this progress on the remainder of the project.

Making Changes

For a schedule to remain viable, the project changes must continually be incorporated into the current schedule. If the update shows that an activity was delayed, extended, interrupted, or accelerated, *P3e* shows the effects of that change on the successors to the activity, and all their successors, all the way to project completion. Before starting the update, you need to save the project baselines. The target dates are the schedule baseline, and the budget amounts are the cost baseline. It is also a good idea to copy your project before updating, for documentation purposes, in case any later problems develop. Generally, a project is updated at the same time each month. The *data date* for recording progress is usually at the beginning of the new month. The current schedule is thus the current plan for project completion. An update should result in a *P3e* project schedule that accurately reflects the current status of the project.

Tracking Physical Progress

There are two phases to keeping the schedule current.

Monitoring Progress. The first is monitoring all activities to determine their status. Each activity is either complete or partially complete, or no work has been accomplished. The activities for which there is no physical progress keep their original activity relationships and durations. Where there is partial physical progress, the original activity relationships are kept and the remaining duration is calculated either as number of days or percent complete. Where activities are complete, the original logic is set aside and the progress is used. After all progress to date is recorded for each activity, the schedule is recalculated. Individual activities and the ensemble are determined to be on, ahead of, or behind schedule.

Documenting Changes. The second phase is documenting changes. Very seldom is a project built in exactly the sequence planned. Either more detail is needed or, with more time for analysis, better plans are developed. When the project scope changes, change orders must be incorporated. As the approach to building the project changes, the contractor needs to modify the schedule to show the changes.

COPYING THE SCHEDULE

P3e offers tools to aid the scheduler in keeping the schedule current. The first step should be to copy the previous schedule. This step is not only in line with good practice but should be considered a requirement. As good practice, one of the advantages of *P3e* is that it allows one to play "what if" games; therefore, for whatever reason, it can be valuable to have an original, "uncontaminated" version of the schedule.

Export Dialog Box.   To make a copy of the schedule, click on the **File** main pull-down menu and select **Export** (Figure 11–1). The **Export** wizard will appear (Figure 11–2). The **Export** wizard, using successive screens, will walk you through the steps of making a file copy of the schedule. Figure 11–2 is the **Export Format** screen of the **Export** wizard. The **Export Format** screen is used to designate the file format of the exported file. The **MPX** format is selected in the **Export Format** screen in Figure 11–2. Figure 11–3 is the **File Name** screen of the **Export** wizard.

ESTABLISHING BASELINE DATES

Before starting the actual update, you need to save the project baseline. The baseline represents the "original" schedule to use for comparison purposes. A baseline is a complete copy of a project plan (in most cases an original schedule) that you can compare to the current schedule to evaluate progress. *P3e* provides you with a number of graphical and

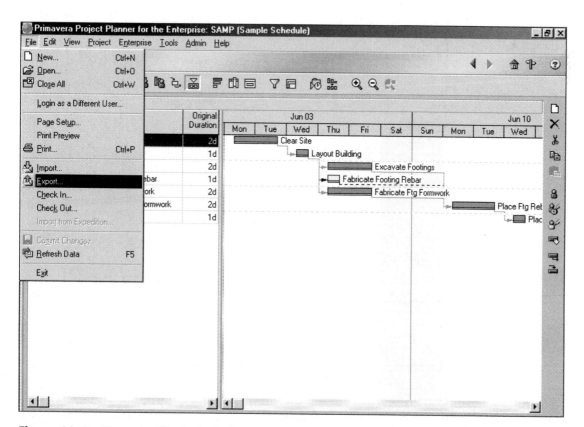

Figure 11–1   Export—File Main Pull-Down Menu

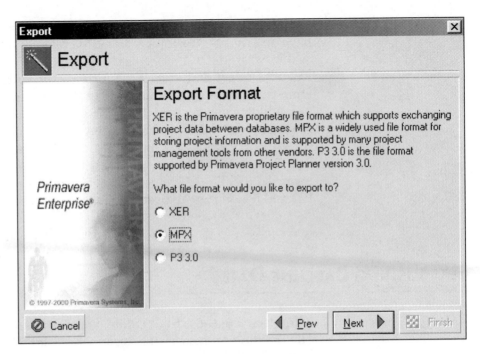

Figure 11-2 **Export** Wizard—**Export Format** Screen

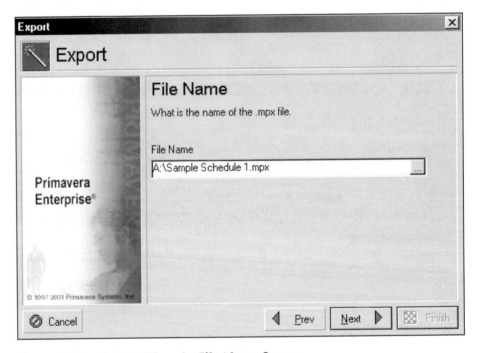

Figure 11-3 **Export** Wizard—**File Name** Screen

tabular layouts for comparing current schedule data with the baseline data. *P3e* has the capability of saving a maximum of 50 baselines per project, but you can select only one baseline as the current baseline for comparison purposes. Baselines do not exist as separate projects that you can access. Therefore, to copy or modify a baseline schedule, you must first unlink it from its current project by restoring it as a separate project. You can then work with this restored baseline project as you would with any other schedule. The fact that in *P3e* a project can have multiple baselines (targets) is extremely important. One baseline could be used to compare the actual to original plan, thus highlighting the overall variance. Another baseline could be used to show the relative improvement or delay from the previous schedule update period.

To create a baseline, select **Baseline** from the **Project** main pull-down menu (Figure 11–1). The **Baselines** dialog box will appear (Figure 11–4). Choose the project, if more than one project is open, and click the **Add** button. The **Add New Baseline** dialog box will appear (Figure 11–5). You can save a copy of the current active project as a new baseline or convert another project into a baseline of the current active project. Note that in Figure 11–5, the **Save a copy of the current project as a new baseline** option is selected and the **OK** button is clicked.

When you save a copy of the current active project as a new baseline, *P3e* titles the new baseline using the project name and a suffix of –Bx, where x equals 1 for the first baseline you save, 2 for the second, and so on. You can change the baseline name. Note that when the **OK** button is clicked in the **Add New Baseline** dialog box (Figure 11–5), you are returned to the **Baselines** dialog box (Figure 11–6). Note that the new baseline project now appears in the dialog box. The current active project is **Sample Schedule**, and a check in the **Use** checkbox of the **Baselines** dialog box makes **Sample Schedule – B1** its baseline schedule.

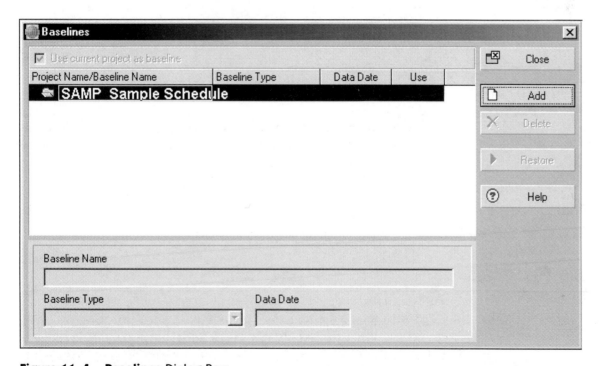

Figure 11–4   Baselines Dialog Box

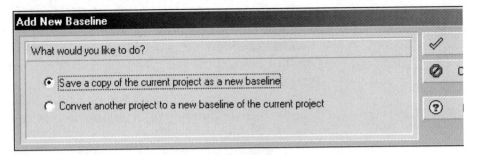

Figure 11–5   Add New Baseline Dialog Box

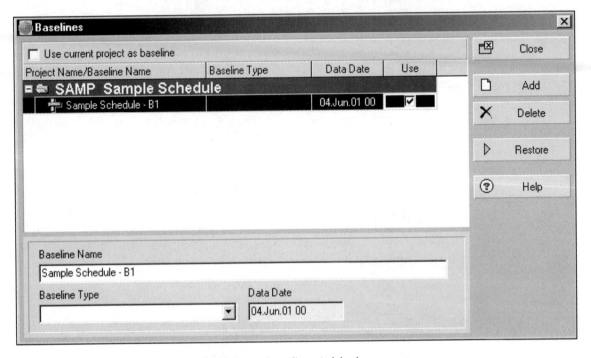

Figure 11–6   Baselines Dialog Box—New Baseline Added

Before converting a project to a baseline, if you still want to have access to the original project, you should make a copy of it. Once you convert a project to a baseline, it is no longer available in the project hierarchy. You can restore a baseline, making it available again as a separate project in the project hierarchy.

RECORD PROGRESS

Progress must be measured before it can be recorded. Physical progress is determined either from daily reports or a mark-up of a tabular schedule report by the project engineer or whomever is responsible for updating the schedule. Table 11–1 is a report showing the actual physical

Activity	Description	Original Duration	Actual Duration	Remaining Duration	% Complete	Actual Start	Actual Finish	Early Finish
A1000	Clear Site	2	3	0	100	4-Jun-01	6-Jun-01	
A1010	Layout Building	1	1	0	100	6-Jun-01	6-Jun-01	
A1020	Excavate Footings	2		1	50	7-Jun-01		11-Jun-01
A1030	Fabricate Ftg Rebar	1	1		100	7-Jun-01	7-Jun-01	
A1040	Fabricate Ftg Formwork	2		1	50	7-Jun-01		11-Jun-01
A1050	Place Ftg Rebar & Formwork	2		2	0			
A1060	Place Ftg Conc	1		1	0			

Table 11–1   Sample Schedule—Actual Physical Progress

progress as determined at the project site for activities A1000, A1010, A1020, A1030, and A1040.

Update Activity Dialog Box

The next step is to record physical progress in *P3e*. Select the first activity for recording progress.

Activity A1000. Choose Activity A1000, Clear Site (Figure 11–7). To record physical progress, select the **Activity Details** button and then the **Status** tab (Figure 11–7). Note that Activity A1000 is highlighted, and the **Status** tab itself has an identifier to show which activity is being updated. When the **Started** checkbox is selected, this means that progress on the activity has actually started. The actual start of Activity A1000 is 04JUN01 according to Table 11–1. Since the planned **Start** (Figure 11–7) and the actual **Start** (Figure 11–8) are the same date, 04JUN01, no change is necessary.

The **Finished** field (Figure 11–7), when checked, becomes the actual **Finish** in Figure 11–8. The actual finish of Activity A1000 is 06JUN01 according to Table 11–1. Since the planned **Finish** (Figure 11–7) and the actual

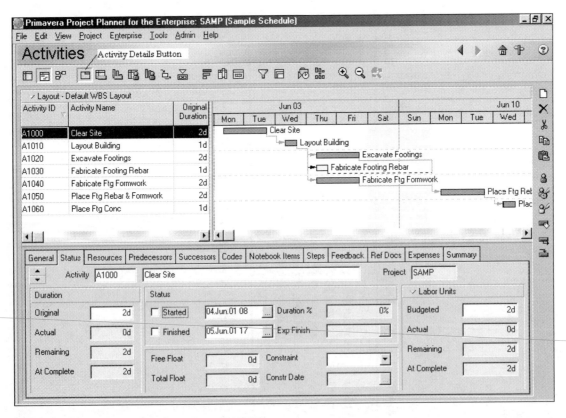

Figure 11–7 **Status** Tab—Activity A1000

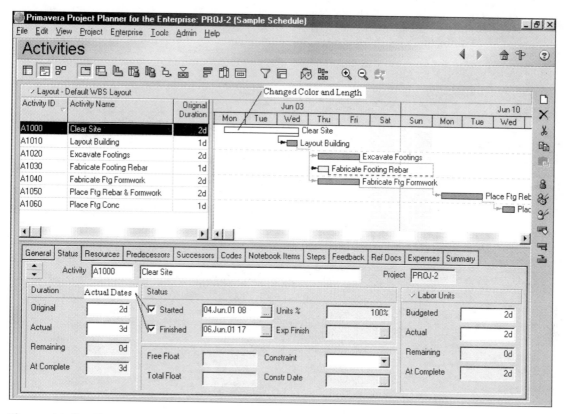

Figure 11–8   Status Tab—Activity A1000—Actual Progress Shown

Finish (Figure 11–8) are not the same date, use the button to expose the pull-down calendar to make the change. Note, in Figure 11–8, that when the **Started** and **Finished** fields are checked to show physical progress, *P3e* automatically changes the **Actual Duration** field to 3d. The **Original Duration** field is unchanged from 2d.

Note that when the **Finished** checkbox is checked in Figure 11–8, the **Remaining** and the **At Complete** fields become grayed out or nonfunctional. This is because when you say the activity is complete, *P3e* automatically sets the remaining duration to 0d and the at completion duration at 3d.

Notice that the activity bar color of Activity A1000 in Figure 11–8 has changed to show progress, and the bar is longer to show the 3-day duration rather than the original 2-day duration. Note that the impact of changing the duration of Activity A1000 did not affect the rest of the activities. The only impact is that Activity A1000 is shown with a 3-day duration rather than a 2-day duration (Figure 11–8).

Activity A1010.   Choose Activity A1010, Layout Building (Figure 11–9). With the **Status** tab still active, click on the next activity in the **Activity Name** field to select it.

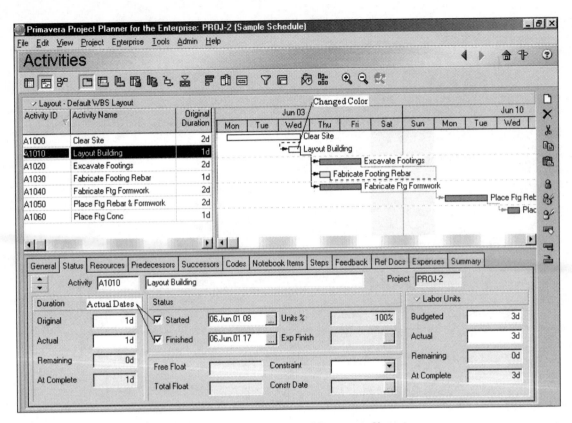

Figure 11–9 **Status** Tab—Activity A1010—Actual Progress Shown

According to Table 11–1, Activity A1010 has an actual duration of 1 day, is 100% complete, and both started and finished on 06JUN01. Since the actual start and finish are the same as planned, simply click in the **Started** and **Finished** checkboxes, and 06JUN01 is automatically accepted as the actual start and finish dates with 0 remaining duration and 100% complete.

Activity A1020. Choose Activity A1020, Excavate Footings (Figure 11–10). With the **Status** tab still active, click on the Activity A1020. According to Table 11–1, Activity A1020 started on 07JUN01, is 50% complete, has 1 day remaining duration, and has a planned early finish of 11JUN01.

Since the activity has been started but not finished, click only on the **Started** checkbox and input 07JUN01. The **Remaining Duration** is set at 1 day. To status activities with actual **Duration**, click the **Remaining** field and enter the remaining number of work periods needed to complete the selected activity. Note that, in Figure 11–10, when the 1 day is placed in the **Remaining** field, the new **Finished** date, the % complete field, and the actual bar length are not automatically calculated by *P3e*. When you schedule or apply actuals (discussed later in this chapter) the **At Complete** duration is calculated as the total working time from the actual

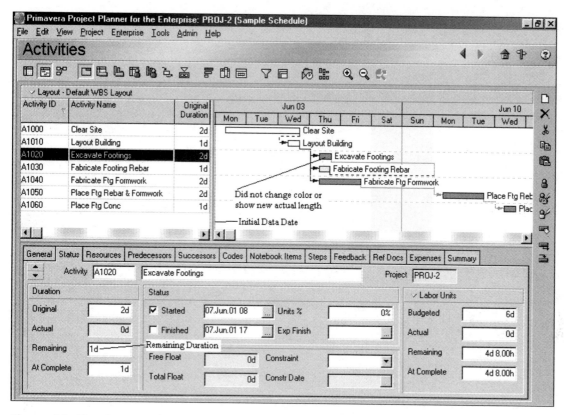

Figure 11–10   Status Tab—Activity A1020—Actual Progress Shown

Started date to the current data date (for the in-progress activities), plus the **Remaining Duration**.

Instead of using the **Remaining Duration** field to claim progress for a partially completed activity, you may use the % **Complete** field. Using the % **Complete** field and the new data date, *P3e* will automatically calculate the new **Remaining Duration** for the partially completed activity.

Activities A1030 and A1040 were updated according to the information in Table 11–1.

APPLY ACTUALS AND ESTABLISH A NEW DATA DATE

Now that progress is input for all the activities where progress is achieved, the next step is to apply the actuals to the schedule as of the new data date. Actuals means the actual physical progress attained of activities that have been started. From the **Tools** main pull-down menu, select **Apply Actuals** and the **Apply Actuals** dialog box appears (Figure 11–11). The **Project(s) to be updated** section lists the **Project Name** for all

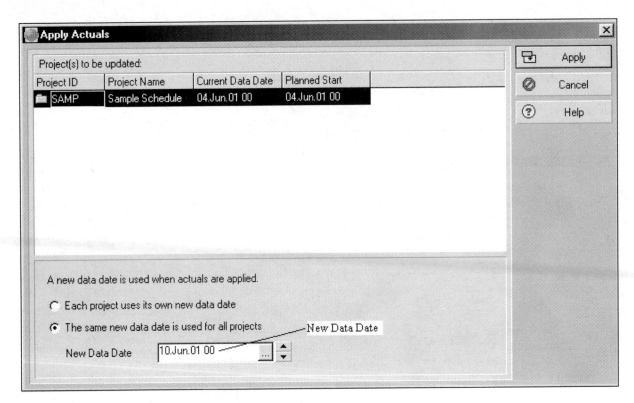

Figure 11–11   Apply Actuals Dialog Box

projects that you can update, along with the **Current Data Date** and the **Planned Start** date for each project. You can specify whether to use the same data date for all projects or whether each project uses its own data date. Note that the **Current Data Date** for the Sample Schedule is 04JUN01 (Figure 11–1).

The data date is the last date for which progress is recorded in the form of actual dates, percentage of work completed, or revised remaining duration. *P3e* schedules activities from the project data date, using the project start date as the initial data date. Common updating data date periods might be at the end of each month, each week, or possibly each project day. Always assume that the data date starts the work period. The data date can be the beginning of the hour, day, week, or month. *P3e* does not credit work accomplished on the data date because it is the beginning of a period.

P3e revises the schedule based on the actual progress you input for the period between the last data date and the current data date. You must change the data date and reschedule a project each time you record progress. During the project, the data date moves forward toward the project completion date. The actual dates can serve as historical data for planning future projects.

Note that in the **New Data Date** field of the **Apply Actuals** dialog box (Figure 11–11), the new data date of 10Jun01 is input. Physical progress is claimed as of the end of the first week of the **Sample Schedule**. Next the **Apply** button of the **Apply Actuals** dialog box is clicked. The Gantt view in Figure 11–12 is the result of the new data date and actuals being applied to the Sample Schedule.

Compare Figure 11–10, before actuals are applied, to Figure 11–12, after the actuals are applied. Note in Figure 11–12, that the data date is moved to 10Jun01. In Figure 11–12, everything to the left of the new data date is a different color from activities to the right. This color distinction and data date location indicate that all activities to the left of the data date are complete or are in progress. All activities to the right of the data date are not complete, or work is still in progress. Note that activity **A1020** now shows (Figure 11–12) the 1d **Remaining Duration** that was input in Figure 11–10 but did not appear in the on-screen Gantt view in Figure 11–10. The work in progress is not shown on the on-screen Gantt view until the progress is claimed using the **Apply Actuals** dialog box and the **New Data Date** is input letting *P3e* know the date to which **Remaining Duration** or % **Complete** applies.

SCHEDULE

Note from Figure 11–12 that activity **A1050** still has the same start date (Figure 11–1) as before progress was entered, actuals were applied, and the new data date was entered. The logic relationships show in Figure 11–12 that activity **A1050** should succeed activities **A1020** and **A1040**. The Sample Schedule must be recalculated for the original logic to be applied to all activities falling after the new data date. The rule of thumb to remember is that all activities falling before the new data date use physical progress and all activities falling after the new data date use the original logic. So, the next step in updating the Sample Schedule is to recalculate it. From the **Tools** main pull-down menu, select **Schedule**. The **Schedule** dialog box will appear (Figure 11–13). The new data date

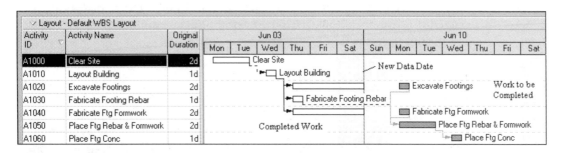

Figure 11–12   Sample Schedule After **Apply Actuals**

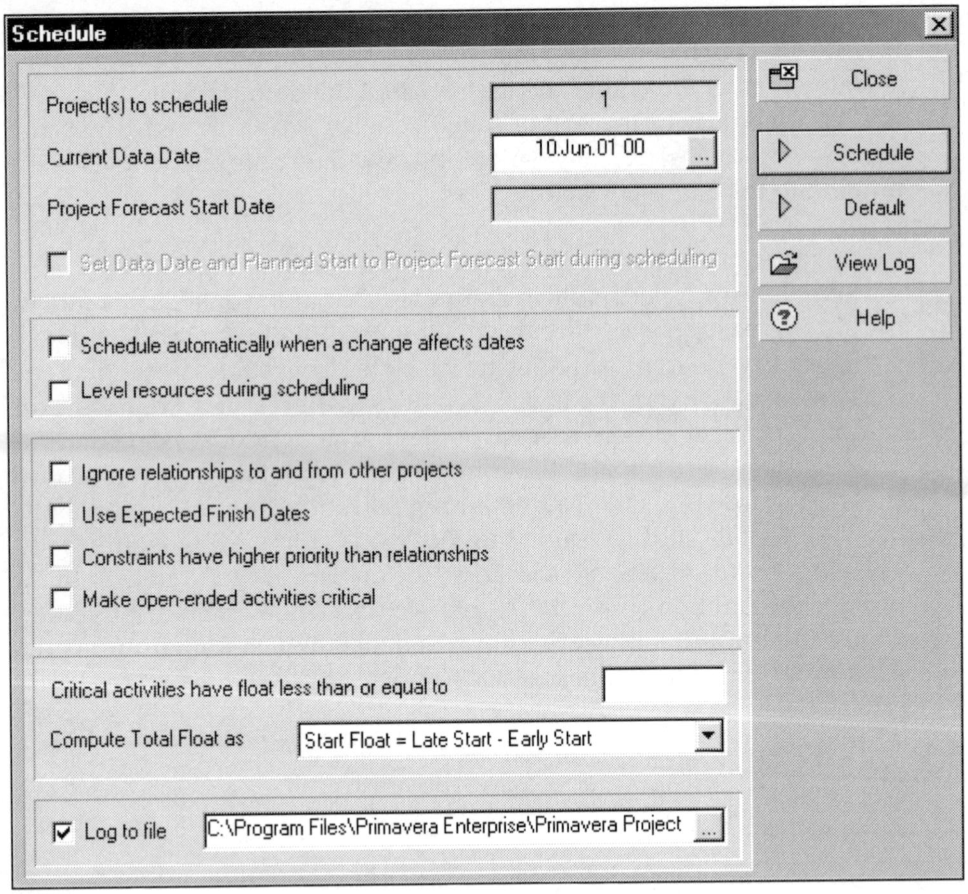

Figure 11–13   Schedule Dialog Box

of 10JUN01 has already been established. Click on the **Schedule** button and *P3e* will perform the forward and backward passes with all calculations. Figure 11–14 is the result of the update on the Sample Schedule. Compare Figure 11–14 to Figure 11–1. The Sample Schedule is essentially one day behind schedule; finishing the project on Thursday rather than Wednesday. Also compare Figure 11–14, after the schedule recalculation, to Figure 11–12 (progress entered, actuals applied, and the new data date entered). Now the correct logic relationships show on the Gantt view. Activity **A1050** succeeds activities **A1020** and **A1040**. *P3e* has used actual progress for activities to the left of the new data date and the original logic input for activities to the right of the new data date.

BASELINE COMPARISON

Seeing a visual comparison of the updated schedule to the baseline dates makes analysis much easier. Click on the **View** main pull-down menu

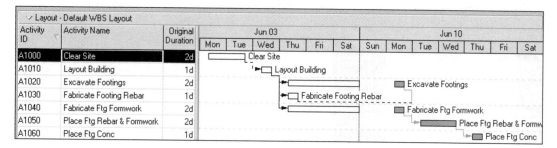

Figure 11–14   Sample Schedule—Results of Schedule Recalculation

and select **Bars**. The **Bars** dialog box will appear (Figure 11–15). Use the button slide bar to scroll to **Baseline**; then in the **Display** column click the checkbox. Compare Figure 11–1 (before the updated baseline schedule) to Figure 11–14 (after the update). Then look at Figure 11–16, which shows both of these schedules on the same screen. For each activity, the top bar represents the current (updated) schedule, and the bottom light bar represents the baseline or original schedule. When you look at Figure 11–16, it is easy to tell where the Sample Schedule fell behind.

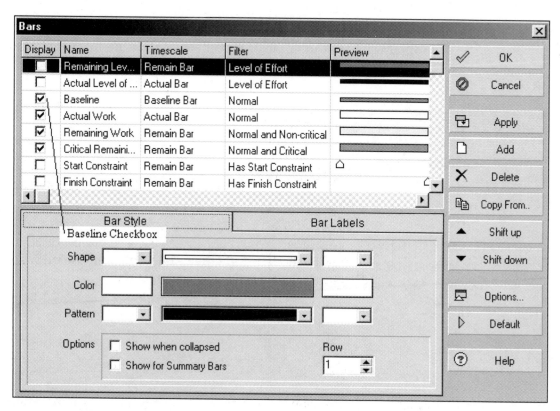

Figure 11–15   Bars Dialog Box

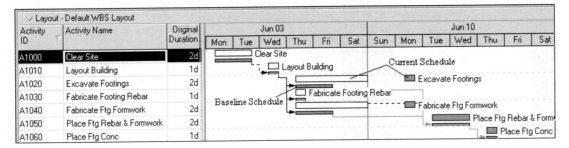

Figure 11–16   Sample Schedule Showing Baseline and Current Schedules

PERT VIEW

To see the result of the update on the *P3e* PERT view, click on the **PERT** button of the Activity Toolbar. The on-screen Gantt chart will convert to the PERT view (Figure 11–17). Notice that the activities that are 100% complete (Activities A1000, A1010, and A1030) show 100% complete. The activities that are partially complete (Activities A1020 and A1040) show 50% complete.

DOCUMENTING CHANGES

As the approach to building the project changes, the schedule should be modified to show the changes. The frequency of updating the schedule

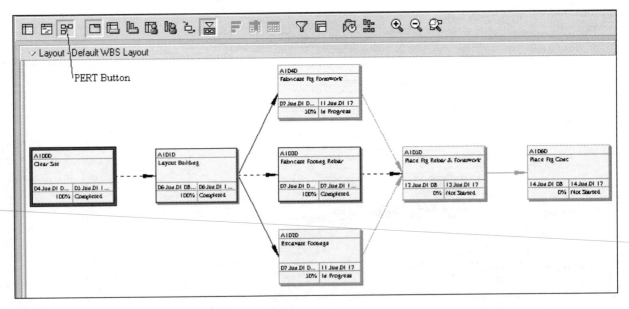

Figure 11–17   PERT View of Updated Schedule

depends on the criticality of the project, the amount of variance, and the contractual conditions. There are many kinds of changes. Activity durations may change if crew sizes are increased to make up for lost time. The logic of activity sequences and interrelationships may be changed to more accurately reflect interferences. Activities may be added or deleted to reflect changes in the scope of the project. Activities may be added for greater detail. As the project proceeds, usually more detailed planning, such as the 10-day look-ahead, is required. As better data become available with actual activity history on which to base projections more details arise. Whatever the source, changes need to be incorporated into the schedule.

Duration Changes

A change in duration may occur because similar activities are performed repeatedly within the schedule. An example is pouring a multi-story concrete building with each floor divided into several pours. By the time several floors have been placed, very good empirical data on the time required for each pour are available. This information is used to adjust projections of activities remaining to be completed.

The original duration of Activity A1050, Place Ftg Rebar & Formwork, was 2 days (see Figure 11–16). In Figure 11–18, it was changed to 3 days. This change was made by changing the number in the **Original Duration** field of the activity table. Notice that when the schedule is recalculated, the impact of this is shown on its successor, Activity A1060. The project duration has been extended from the end of Thursday to Friday.

Logic Changes

The original logic of the relationships between activities may change for any number of reasons, such as:

• Adding missing activities
• Removing redundant activities
• Modifying logic in case of progress overide

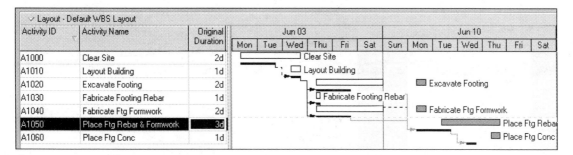

Figure 11–18   Sample Schedule—Changed Duration

The following is an example of a logic change and how to incorporate it into the schedule. It is necessary for inspection purposes to include one day of lag between Activities A1050 and A1060. Select Activity A1050; then click the **Activity Details Button**. Then select the **Successions** tab (Figure 11–19). Activity A1060 is a **Successor** to Activity A1050. Select Activity A1050 for changes; then click on the **Lag** field and enter 1. Compare the calculated schedule with this change incorporated (Figure 11–19) to the schedule before the change (Figure 11–18). In the schedule before the change, Activity A1050 ends at the end of Thursday. Activity A1060 ends at the end of Friday. With the addition of the lag day in Activity A1050, it ends at the end of Thursday. Friday is the lag day. Saturday and Sunday are nonworkdays according to the project calendar configuration, so Activity A1060 now begins and ends on Monday.

Added Activities

Sometimes it is necessary to either add or delete activities from the schedule. Adding activities to the schedule within *P3e* is not difficult. Place the mouse cursor on the activity after which the new activity will be placed (Figure 11–20). Then, to add the new activity, there are two approaches within *P3e*. The first approach is to click the **Add** button from the Gantt screen and you will activate the *P3e*'s **Add Activity Wizard**.

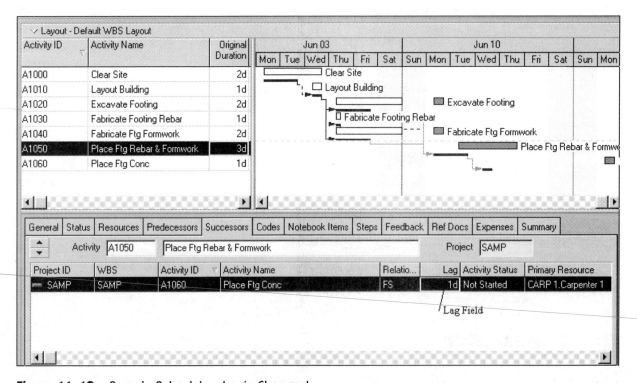

Figure 11–19   Sample Schedule—Logic Changed

The second approach is to select **Add** from the **Edit** main-pull down menu. The second method was used to create the new activity in Figure 11–20. Note in Figure 11–20 that **Activity Details** has been activated for adding information about the new activity.

In Figure 11–20, the new activity, A1070, Strip, Clean & Oil Forms, with an original duration of 1 day, is placed after the last activity, A1060, Place Ftg Conc. In Figure 11–20, *P3e* places the new activity after the data date. This is because we have not yet established relationships with the other activities, and this is the earliest possible time that the activity could be started, since the time before the data date has been expended.

Relationships.   Relationships need to be defined next. Since Activity A1070 is the last activity of the schedule and will have no successors, predecessors need to be established. Highlight Activity A1070, and then click the **Predecessors** tab. In Figure 11–21, Activity A1060, Place Ftg Conc, is made a predecessor to Activity A1070, Strip, Clean & Oil Forms. The default **Relationship** type of Finish-to-Start (**FS**), and 0 **Lag** time is accepted. Figure 11–21 is the schedule recalculated with the new activity. The new planned project finish date is Tuesday.

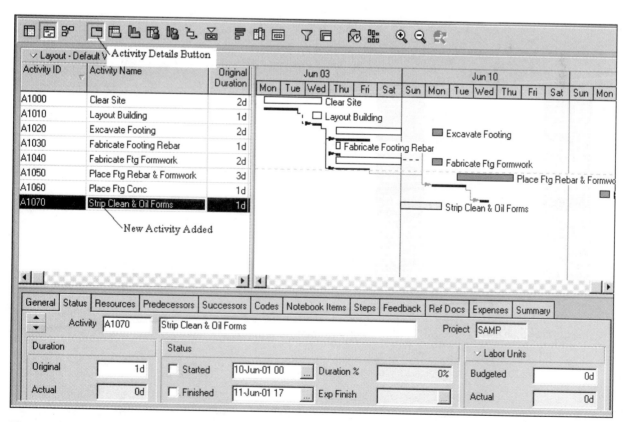

Figure 11–20   Sample Schedule—New Activity Added

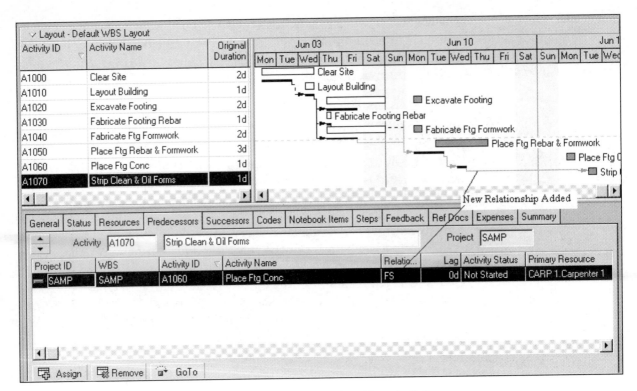

Figure 11–21 Sample Schedule—Relationship Added to New Activity

EXAMPLE PROBLEM

Table 11–2 was used to update the schedule for the house put together as an example for student use (see the wood-framed house drawings in the Appendix). Figure 11–22 is a hard-copy print of the updated *P3e* tabular report. Figure 11–23 is a hard-copy print of the updated Gantt chart comparison report.

Activity ID	Activity Description	Original Duration	Actual Duration	Remaining Duration	% Complete	Actual Start	Actual Finish
A1000	Clear Site	2	3	0	100	4-Sep-01	6-Sep-01
A1010	Building Layout	1	1	0	100	4-Sep-01	4-Sep-01
A1020	Form/Pour Footings	3	3	0	100	5-Sep-01	7-Sep-01
A1030	Pier Masonry	2	1	0	100	10-Sep-01	10-Sep-01
A1040	Wood Floor System	4	3	0	100	11-Sep-01	13-Sep-01
A1050	Rough Framing Walls	6		4	33	14-Sep-01	
A1060	Rough Framing Roof	4			0		
A1070	Doors & Windows	4			0		
A1080	Ext Wall Board	2			0		
A1090	Ext Wall Insulation	1			0		
A1100	Rough Plumbing	4			0		
A1110	Rough HVAC	3			0		
A1120	Rough Elect	3			0		
A1130	Shingles	3			0		
A1140	Ext Siding	3			0		
A1150	Ext Finish Carpentry	2			0		
A1160	Hang Drywall	4			0		
A1170	Finish Drywall	4			0		
A1180	Cabinets	2			0		
A1190	Ext Paint	3			0		
A1200	Int Finish Carpentry	4			0		
A1210	Int Paint	3			0		
A1220	Finish Plumbing	2			0		
A1230	Finish HVAC	3			0		
A1240	Finish Elect	2			0		
A1250	Flooring	3			0		
A1260	Grading & Landscaping	4			0		
A1270	Punch List	2			0		

Table 11–2   Wood-Framed House—Actual Physical Progress

Wood Frame House
Report Date 28-Jul-01 09

Start Date 04-Sep-01
Finish Date 15-Nov-01
Data Date 21-Sep-01

SR-02 Schedule Report Comparison to Target

WBS

Activity ID	Activity Name	At Comp Dur	BL Dur	Activity % Comp	Start	Finish
Example	**Wood Frame House**					
A1000	Clear Site	3	2	100%	04-Sep-01 08 A	06-Sep-01 17 A
A1010	Building Layout	1	1	100%	04-Sep-01 08 A	04-Sep-01 17 A
A1020	Form/Pour Footings	3	3	100%	05-Sep-01 08 A	07-Sep-01 17 A
A1030	Pier Masonry	1	2	100%	10-Sep-01 08 A	10-Sep-01 17 A
A1040	Wood Floor System	3	4	100%	11-Sep-01 08 A	13-Sep-01 17 A
A1050	Rough Framing Walls	10	6	33.33%	14-Sep-01 08 A	27-Sep-01 17
A1060	Rough Framing Roof	4	4	0%	27-Sep-01 08	02-Oct-01 17
A1070	Doors & Windows	4	4	0%	05-Oct-01 08	10-Oct-01 17
A1080	Ext Wall Board	2	2	0%	03-Oct-01 08	04-Oct-01 17
A1090	Ext Wall Insulation	1	1	0%	16-Oct-01 08	16-Oct-01 17
A1100	Rough Plumbing	4	4	0%	05-Oct-01 08	10-Oct-01 17
A1110	Rough HVAC	3	3	0%	05-Oct-01 08	09-Oct-01 17
A1120	Rough Elect	3	3	0%	10-Oct-01 08	12-Oct-01 17
A1130	Shingles	3	3	0%	11-Oct-01 08	15-Oct-01 17
A1140	Ext Siding	3	3	0%	18-Oct-01 08	22-Oct-01 17
A1150	Ext Finish Carpentry	2	2	0%	16-Oct-01 08	17-Oct-01 17
A1160	Hang Drywall	4	4	0%	17-Oct-01 08	22-Oct-01 17
A1170	Finish Drywall	4	4	0%	23-Oct-01 08	26-Oct-01 17
A1180	Cabinets	2	2	0%	29-Oct-01 08	30-Oct-01 17
A1190	Ext Paint	3	3	0%	23-Oct-01 08	25-Oct-01 17
A1200	Int Finish Carpentry	4	4	0%	31-Oct-01 08	05-Nov-01 17
A1210	Int Paint	3	3	0%	06-Nov-01 08	08-Nov-01 17
A1220	Finish Plumbing	2	2	0%	31-Oct-01 08	01-Nov-01 17
A1230	Finish HVAC	3	3	0%	29-Oct-01 08	31-Oct-01 17

Page 1 of 2

Figure 11–22   Updated Tabular Report—Wood-Framed House

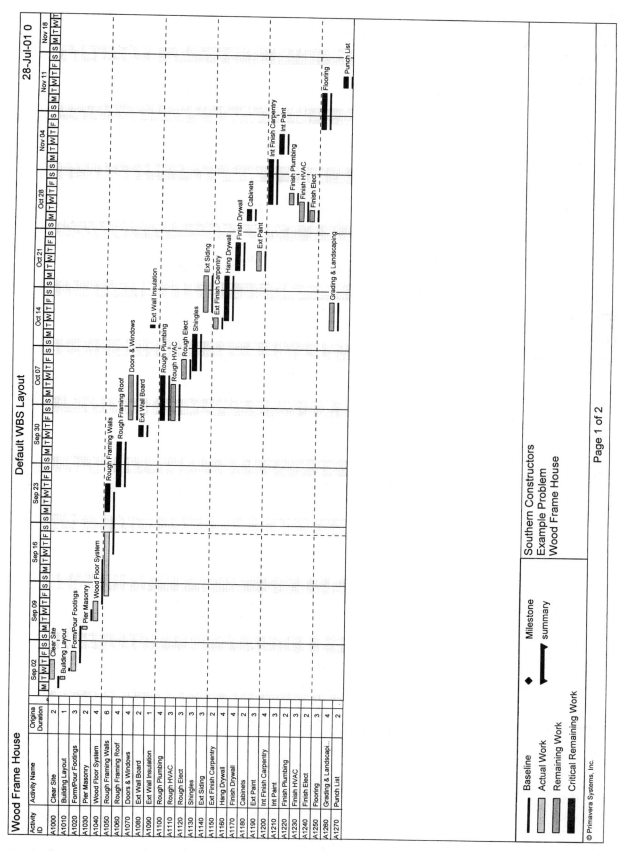

Figure 11–23   Updated Gantt Chart—Wood-Framed House

Exercises

1. Update Figure 11–24 manually; then update it on-screen using the *P3e* Gantt chart.

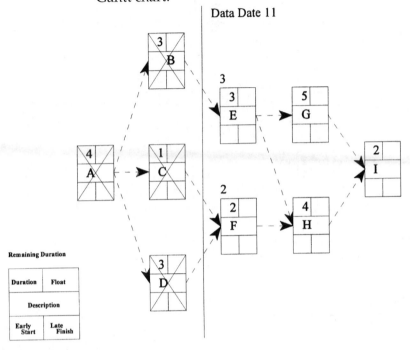

Figure 11–24   Exercise 1

2. Update Figure 11–25 manually; then update it on-screen using the *P3e* Gantt chart.

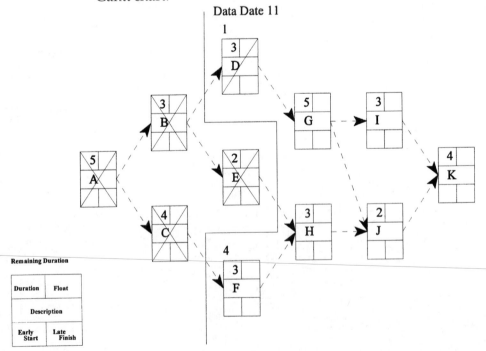

Figure 11–25   Exercise 2

3. Update Figure 11–26 manually; then update it on-screen using the *P3e* Gantt chart.

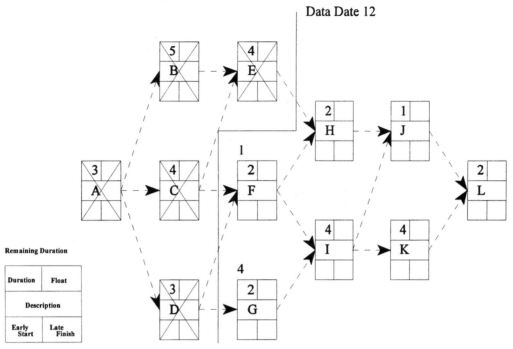

Figure 11–26   Exercise 3

4. Update Figure 11–27 manually; then update it on-screen using the *P3e* Gantt chart.

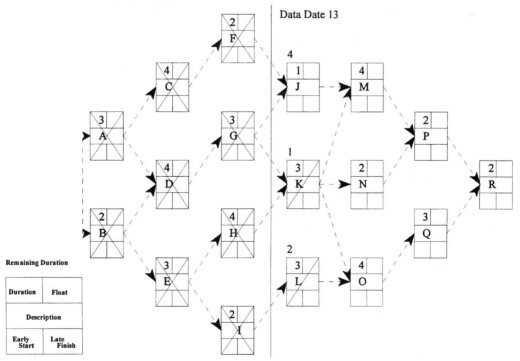

Figure 11–27   Exercise 4

5. Update Figure 11–28 manually; then update it on-screen using the *P3e* Gantt chart.

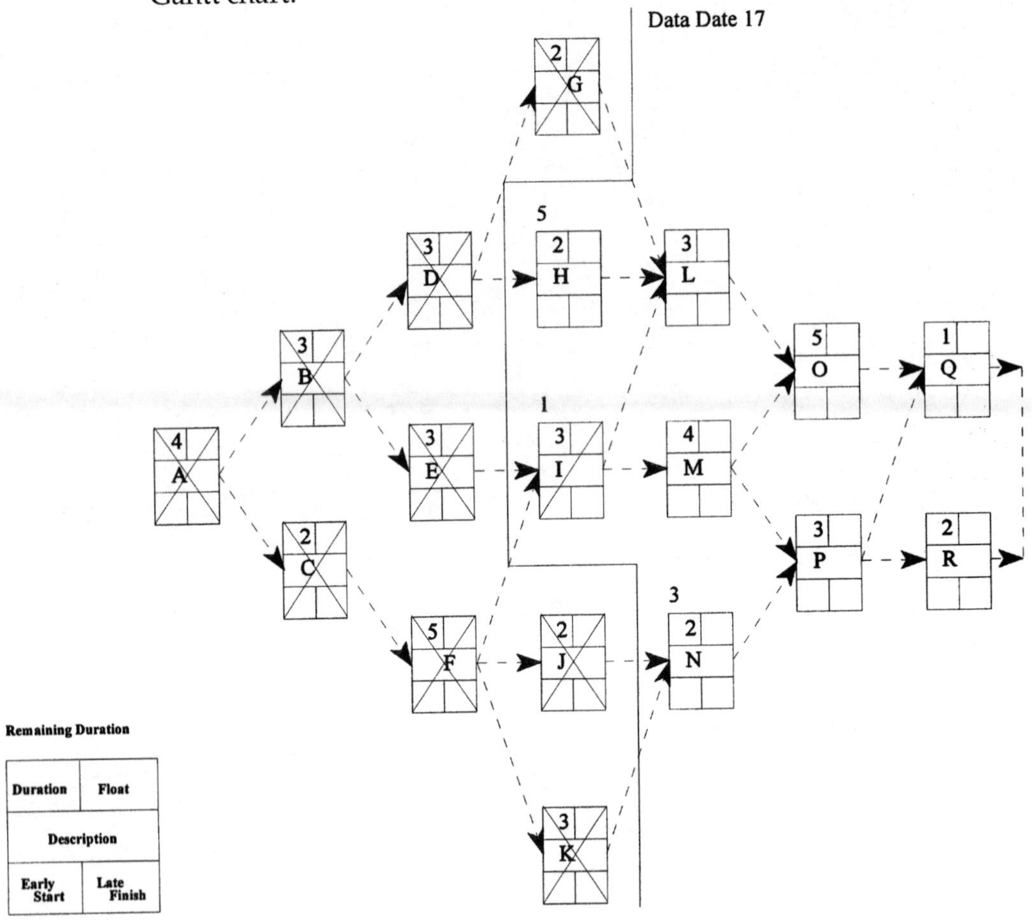

Figure 11–28 Exercise 5

6. Update Figure 11–29 manually; then update it on-screen using the *P3e* Gantt chart.

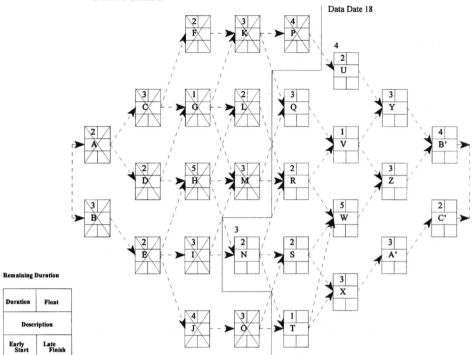

Figure 11-29   Exercise 6

7. Small Commercial Concrete Block Building—Updating
Prepare the following reports for the small commercial concrete block building located in the Appendix.
A. Updated Gantt chart.
B. Schedule report comparison to target in tabular format.

8. Large Commercial Building—Updating
Prepare the following reports for the large commercial building located in the Appendix.
A. Updated Gantt chart.
B. Schedule report comparison to target in tabular format.

12

Tracking Resources

Objectives

Upon completion of this chapter, you should be able to:

- Record resource usage
- Use updated resource spreadsheets
- Use updated resource profiles
- Use updated resource tabular reports

TRACKING RESOURCES: ACTUAL VERSUS PLANNED EXPENDITURES

Once a project is under way, the contractor tracks (monitors) resources to compare actual to baseline progress. The resources can be labor, equipment, materials, or other resources that were assigned to the activity (usually allocated from the original estimate) when the baseline schedule was constructed. Figure 12–1 shows the Sample Schedule in the **Gantt Chart** view with the **Activity Usage Spreadsheet** appearing at the bottom of the screen. The **Gantt Chart** view at the top of the screen shows the baseline version of the Sample Schedule. The **Activity Usage Spreadsheet** window is obtained by selecting the **Activity Usage Spreadsheet** button from the **Activity Toolbar** (Figure 12–1). The **Activity Usage Spreadsheet** is configured to show the **Budgeted Labor Units**, which correspond to the baseline version of the Sample Schedule. Table 12–1

	Mon	Tue	Wed	Thu	Fri	Sat	Sun	Mon	Tue	Wed
CARP 1	1000-1	1000-1	1010-1	1020-1 1040-1	1020-1 1040-1			1050-1	1050-1	1060-1
CARP 2			1010-1	1040-1	1040-1					
EQUIP OP				1020-1	1020-1			1050-1	1050-1	1060-1
LAB 1			1010-1	1020-1 1030-1	1020-1					1060-1
LAB 2				1030-1 1040-1	1040-1			1050-1	1050-1	
Total	1	1	3	8	6	0	0	3	3	3

Table 12–1 Craft Requirements by Activity by Day—Baseline Schedule

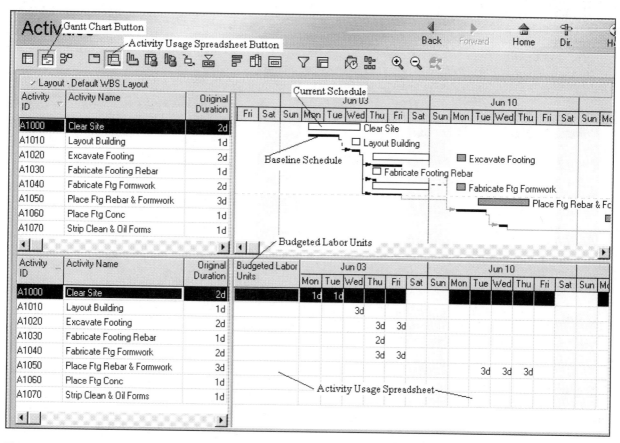

Figure 12–1   Sample Schedule Showing **Activity Usage Spreadsheet**

shows the detailed craft requirements sorted by activity by day for the baseline (original) schedule.

In Chapter 11, we updated the project for current progress. This updated schedule is now called the current schedule. It can be compared to the baseline schedule to determine actual physical progress. Figure 12–1 shows the current version of the Sample Schedule as updated in Chapter 11. The **Activity Usage Spreadsheet** needs to be reconfigured to also show the **Actual Labor Units** expended for activities with actual physical progress as updated in Chapter 11. To accomplish the reconfiguration of the **Activity Usage Spreadsheet**, click the right mouse button while in the **Activity Usage Spreadsheet**. Select **Spreadsheet Fields**, and the **Fields** dialog box will appear (Figure 12–2). Under **Time Interval Units**, use the directional arrows to move **Actual Labor Units** to the **Selected Options** field. Figure 12–3 has rows for both the **Actual** and **Budgeted Labor Units** appearing in the **Activity Usage Spreadsheet**. Now making a comparison between the planned and actual expenditure of resources is much simpler.

Only by tracking and comparing the actual usage of resources to the baseline budget can the contractor determine physical progress and

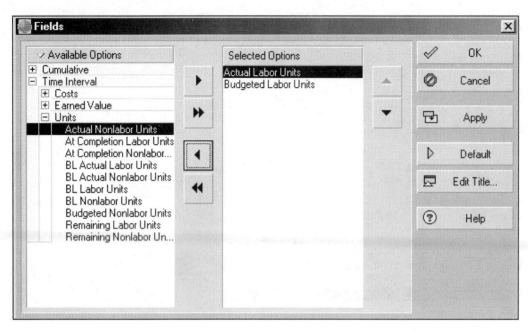

Figure 12–2   Fields Dialog Box

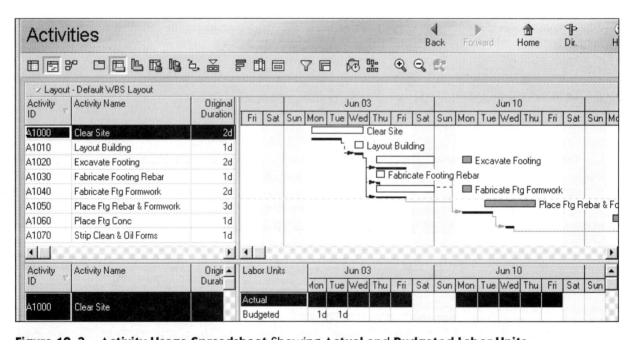

Figure 12–3   Activity Usage Spreadsheet Showing **Actual** and **Budgeted Labor Units**

earned value. Management must gauge the progress gained for the amount of resources expended to determine if the project is over or under the baseline budget. This analysis provides the information to help solve potential cost and time problems.

On a cost-plus or negotiated project, the owner has access to the contractor's resource information. The resource-loaded and then resource-monitored schedule thus gives the owner valuable information about how the contractor is managing his or her company's time and cash.

RECORD EXPENDITURES

The labor resources as input in Chapter 6 are tracked in this section. The schedule will be updated to show the actual expenditures of resources and to make forecasts.

Activity A1000

The recording of the actual expenditure of resources for Activity A1000 will be input in the following sections.

Activity Details. On the *P3e* Gantt chart screen, click on the activity to be updated. Then, from the Activity Toolbar, select the **Activity Details** button. In Figure 12–4, Activity A1000 has been selected. Look at the

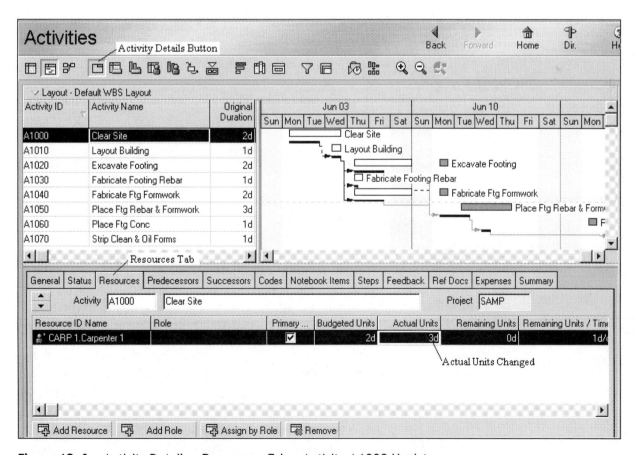

Figure 12–4   Activity Details—**Resources** Tab—Activity A1000 Update

Resources tab window at the bottom of the screen to check the updated information. The original budgeted units are 2d (days).

In Figure 12–3, the baseline schedule shows a 2-day duration, with one CARP 1 for Mon, and one CARP 1 for Tue. When the schedule was updated in Chapter 11, the actual duration for Activity A1000 was changed to 3 days.

Actual Labor Expenditures. From project time sheets, the actual labor expenditures and projections to completion are gathered (Table 12–2). For Activity A1000, 1 person for 3 days was actually used, rather than 1 person for 2 days; therefore, a total of 3 days of CARP1 time was expended. The **Resources** tab for Activity A1000 (Figure 12–4) must be updated to show actual progress. In Figure 12–4, the **Actual Units** field for CARP 1 is changed from the *P3e*-generated 2d to 3d.

When this change is made (Figure 12–4), the **Budgeted Units** field remains at 2.00, since the baseline budget has not changed.

Activity Usage Spreadsheet. Compare the **Activity Usage Spreadsheet** before the resource update for Activity A1000 (Figure 12–3) with that after the update (Figure 12–5). The **Actual Labor Units** row in Figure 12–5, the **Activity Usage Spreadsheet** after the update, now shows 1d of CARP 1 labor expended for Mon, Tue, and Wed.

Activity A1010

The recording of the actual expenditure of resources for Activity A1010 will be input in the following sections.

Activity Details. Click on the next activity to be updated, Activity A1010. Look at the **Resources** tab of the **Activity Details** window (Figure 12–6) to check the updated information. The **Budgeted Units** is 1 day.

Actual Labor Expenditures. When the schedule was updated in Chapter 11, the actual duration for Activity A1010 was 1 day (Mon). The **Budgeted Units** of 1 for CARP 1 and CARP 2 was correct according to cost information received (Table 12–2), but 2 person-days for LAB 1 was expended, so the **Actual Units** field for Activity A1010 must be updated to show actual progress. In Figure 12–6, the **Actual Units** field for LAB 1 is changed from the *P3e*-generated 1d to 2d. The **Actual Units** field for CARP 1 and CARP 2 are left unchanged at the *P3e*-generated 1d.

Activity Usage Spreadsheet. When the updated resource requirements for this activity are input, the new **Activity Usage Spreadsheet** window shows the actual expenditures. The **Actual Labor Units** row in Figure 12–7, the **Activity Usage Spreadsheet** after the update, now shows 4d expended for Wed (1d for CARP 1, 1d for CARP 2, and 2d for

Activity	Percent Complete	Actual Expenditures to Date (Person-days)					Projected Expenditures to Completion				
		Carp 1	Carp 2	Equip Op	Lab 1	Lab 2	Carp 1	Carp 2	Equip Op	Lab 1	Lab 2
A1000	100	3									
A1010	100	1	1								
A1020	50	1		1	2		1		1	3	
A1030	100				2	2					
A1040	50	1	1			1	1	1			1
A1050	0						3		3		3
A1060	0						1		1	1	
A1070	0								1	1	

Table 12–2   Craft Requirements—Actual Expenditures to Date and Projected Expenditures to Completion

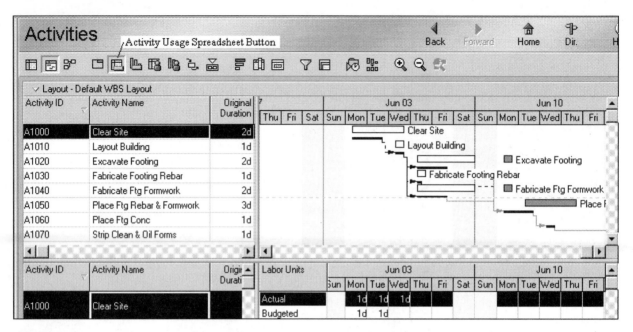

Figure 12–5   Activity Usage Spreadsheet—Activity A1000 Updated

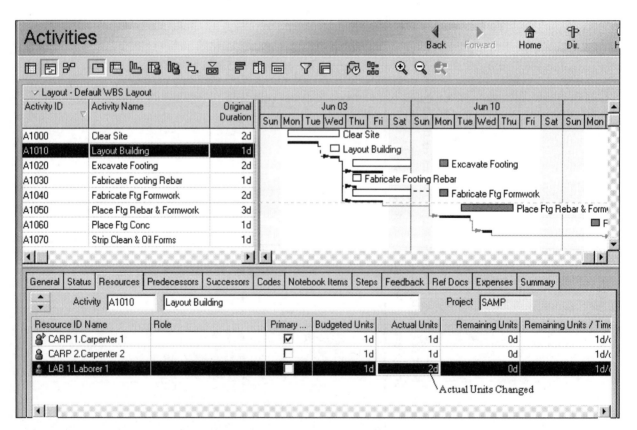

Figure 12–6   Activity Details—Resources Tab—Activity A1010 Update

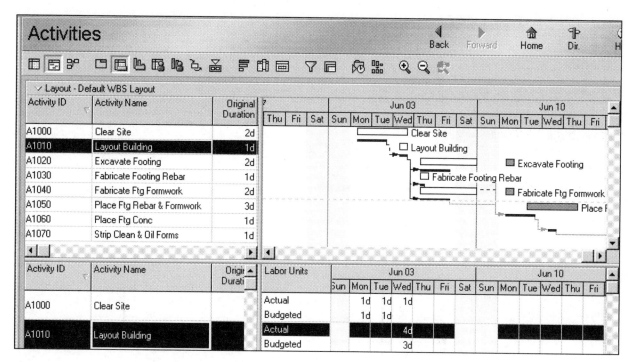

Figure 12-7   Activity Usage Spreadsheet—Activity A1010 Updated

LAB 1). The **Budgeted Labor Units** row still shows the baseline original budget of 3d (1d for CARP 1, 1d for CARP 2, and 1d for LAB 1).

Activity A1020

The recording of the actual expenditure of resources for Activity A1020 will be input in the following sections.

Activity Details.   Click on the next activity to be updated, Activity A1020. Look at the **Resources** tab of the **Activity Details** window (Figure 12–8) to check the updated information. The **Budgeted Units** are 6 days (2d for CARP 1, 2d for EQUIP OP, and 2d for LAB 2).

Actual Labor Expenditures.   The **Gantt Chart** shown in Figure 12–8 is changed to reveal the **Remaining Duration** and **Performance % Complete** columns. Note that the **Original Duration** of Activity A1020 is 2d. The **Remaining Duration** is 1d and the **Performance % Complete** is 50%. The **Budgeted Units** (from the **Resources** tab of the **Activity Details** in Figure 12–8) are a total of 6d (2d for CARP 1, 2d for EQUIP OP, and 2d for LAB 2). From the actual expenditure of resource information (Table 12–2), the **Budgeted Units** for the CARP 1 and EQUIP OP in the

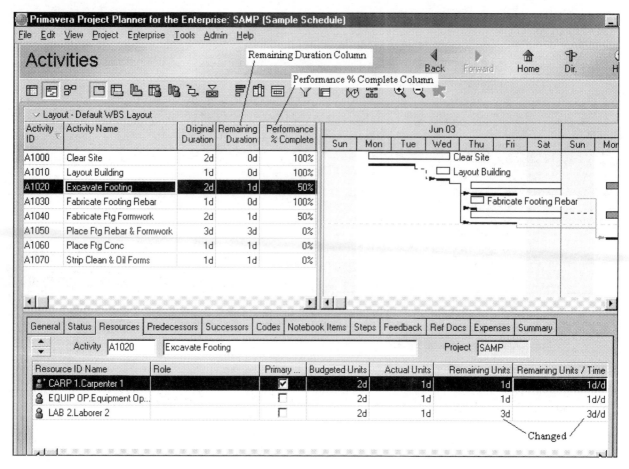

Figure 12–8 Activity Details—Resources Tab—Activity A1020 Update

baseline schedule were correct. Fifty percent of both budgets have been expended (1d of **Actual Units**) and 50% remain to be expended (1d of **Remaining Units**). The **Budgeted Units** for the LAB 2 in the baseline schedule were not correct. Fifty percent of the budget has been expended (1d of **Actual Units**) but the **Remaining Units** in Figure 12–8 had to be changed to reflect the 3d remaining. The **Remaining Units/Time** field in Figure 12–8 also had to be changed to reflect the 3d/d number of units remaining to be used.

Activity Usage Spreadsheet. When the updated resource requirements for this activity are input, the new **Activity Usage Spreadsheet** is shown in Figure 12–9. The rows in the **Activity Usage Spreadsheet** have been modified to show the **Remaining Labor Units** row along with the **Actual** and **Budgeted Labor Units** rows. Note that the **Budgeted Labor Units** have 3d for Thu and Fri as input in the baseline schedule. The **Actual Labor Units** now shows 2d for Thu and Fri. The actual requirements from Figure 12–8 are 1d for CARP 1, 1d for EQUIP OP, and 1d for LAB 2 or a total of three days actually expended. *P3e* takes the three days of labor time expended and divides it by the two days duration to get

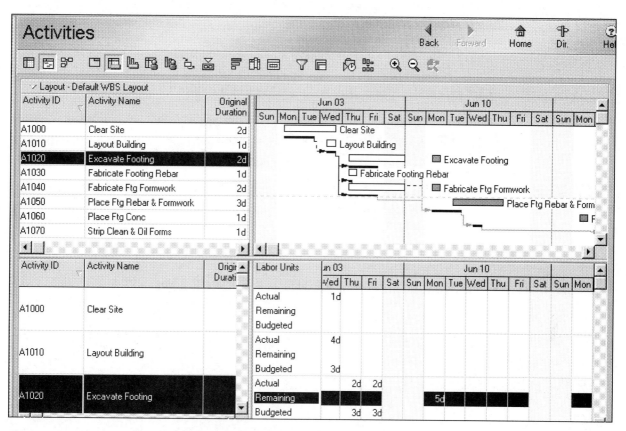

Figure 12–9   Activity Usage Spreadsheet—Activity A1020 Updated

one and one-half days of labor time expended/duration day. *P3e* then rounds up to the nearest whole number or a 2d average labor time expenditure.

The **Remaining Labor Units** show 5d for Mon. The remaining requirements from Figure 12–8 are 1d for CARP 1, 1d for EQUIP OP, and 3d for LAB 2 or a total of five days.

Activities A1030 and A1040

The input of the actual expenditure of resources is input into the Sample Schedule according to that shown in Table 12–2. The same procedures were used in inputting the expenditure of resources in Activities A1030 and A1040 as with Activities A1000, A1010, and A1020.

Activities A1050 and A1060

Since there is no actual progress or expenditure of resources shown in Table 12–2 for Activities A1050 and A1060, there are no changes made to the **Resources** tab of the **Activity Details** for these two activities.

Activity 1070

Click on the last activity to be updated, Activity A1070. This activity was added in the update to the Sample Schedule in Chapter 11. The original duration shown is 1. The percent complete is 0%.

Input Budget.   In Chapter 11, no resources were added when Activity A1070 was added to the schedule. According to Table 12–2, the following changes need to be made to the **Resources** tab of the **Activity Details**.

1. **Resources:** Add resources EQUIP OP and LAB 1 (Figure 12–10).
2. **Budgeted Units:** Add **Budgeted Units** of 1d for EQUIP OP and 2d for LAB 1.

OTHER *P3E* WINDOWS FOR ANALYZING RESOURCE UPDATES

So far in this chapter, only the **Activity Usage Spreadsheet** window (Figures 12–2, 12–3, 12–5, 12–7, and 12–9) has been used to analyze the actual

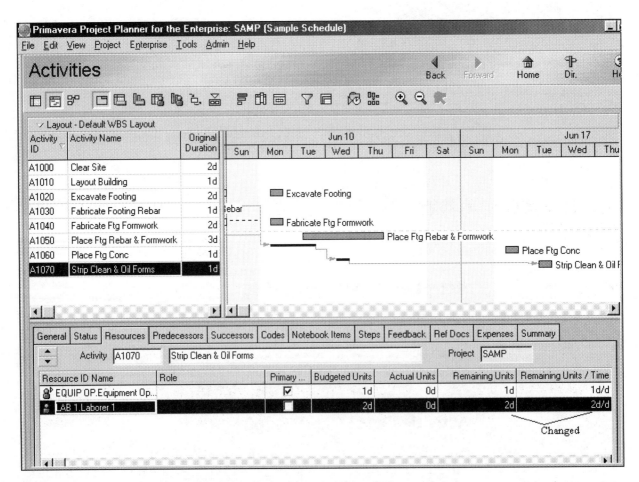

Figure 12–10   Activity Details—Resources Tab—Activity 1070

expenditure of resources in the Sample Schedule. *P3e*'s other views also offer valuable tools for analysis of resource expenditures. Other *P3e* windows (**Resource Usage Spreadsheet**, **Activity Usage Profile**, and **Resource Usage Profile**) also offer powerful analytical capability.

Resource Usage Spreadsheet.   The **Resource Usage Spreadsheet** is obtained by clicking the **Resource Usage Spreadsheet** button from the Activity Toolbar (Figure 12–11). Note in Figure 12–11 that the resource CARP 1 is selected in the **Resource ID** field. The first row of the spreadsheet portion of the **Resource Usage Spreadsheet** shows the total requirements for the selected resource by time period. The **Resource Usage Spreadsheet** also shows all activities where CARP 1 (the selected resource) is specified along with the **Actual**, **Remaining**, and **Budgeted** labor units. This window is valuable for analyzing individual resources by activity.

If costs rather than resources appear when the **Resource Usage Spreadsheet** is selected, you must reconfigure the spreadsheet. While in the **Activity Usage Profile**, click the right mouse button and select **Activity Usage Profile Options** to reconfigure the profile. When you return to the spreadsheet, the reconfiguration will still be in effect.

Activity Usage Profile.   The **Activity Usage Profile** is obtained by clicking the **Activity Usage Profile** button from the Activity Toolbar

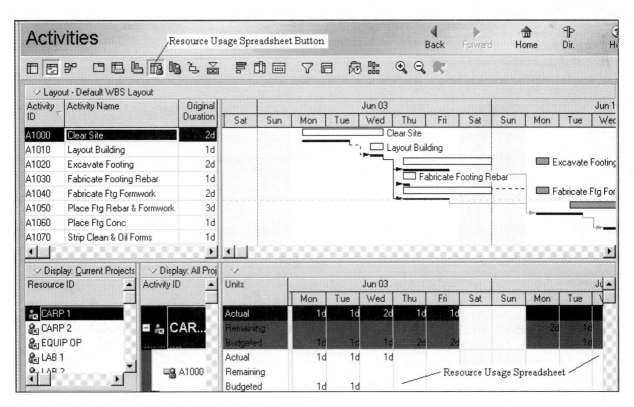

Figure 12–11   Resource Usage Spreadsheet Window

(Figure 12–12). Note in Figure 12–12 that the profile is used to determine the resource totals for **All Activities** or for **Selected Activities**. Figure 12–12 shows the resource totals for all activities. This window is valuable for the analysis of the overall labor unit requirements for the project. The **Activity Usage Profile** is valuable in planning project labor/resource requirements.

If costs rather than resources appear when the **Activity Usage Profile** is selected, you must reconfigure the profile. While in the **Activity Usage Profile**, click the right mouse button and select **Activity Usage Profile Options** to reconfigure the profile.

Resource Usage Profile. The **Resource Usage Profile** is obtained by clicking the **Resource Usage Profile** button from the Activity Toolbar (Figure 12–13). Note in Figure 12–13 that the profile is used to determine the individual resource totals. Figure 12–13 shows the resource profile for CARP 1. This window is valuable for the analysis of individual resource/labor units requirements for the project.

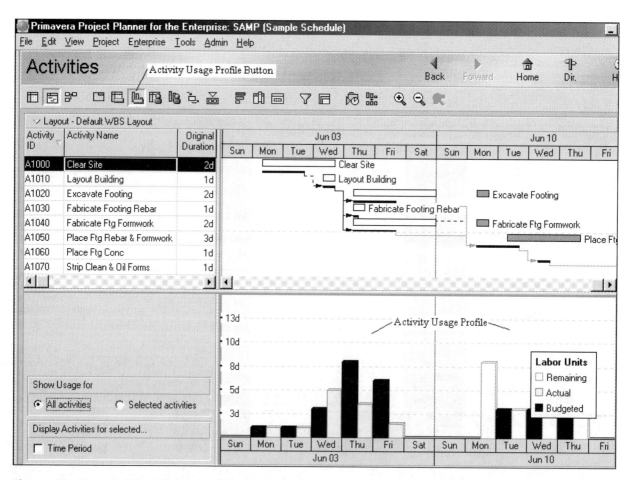

Figure 12–12   Activity Usage Profile Window

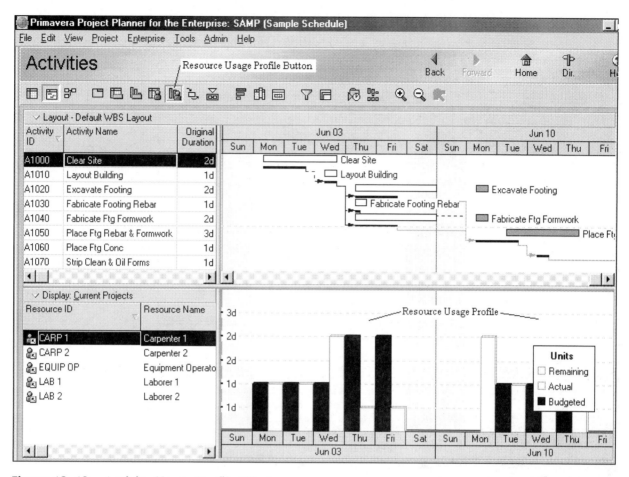

Figure 12–13   Activity Usage Profile Window

RESOURCE REPORTS

Figure 12–14 is the **Print Preview** of the Sample Schedule in Gantt chart format showing the **Activity Usage Profile**. All resources to the left of the data date are actual to-date entries. All entries to the right of the data date are to-complete estimates. Figure 12–15 is the **Print Preview** of the **Resource Usage Profile**. Figure 12–16 is the **Print Preview** of the **Resource Usage Spreadsheet**.

Figure 12–17 shows what can be done with *P3e*'s customized tabular reports. This comparison report (**Budgeted Units, Actual Units, Remaining Units** versus **At Completion Units**) is created using the **Reports** selection from the **Tools** main pull-down menu.

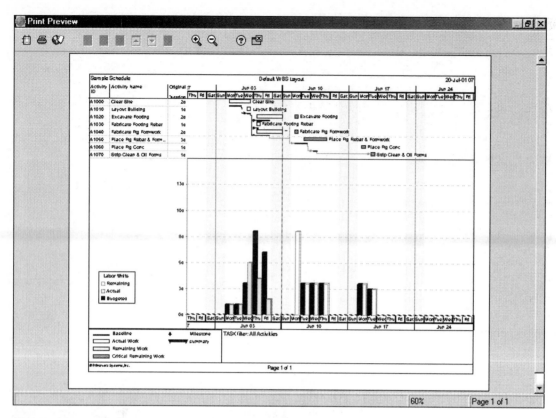

Figure 12-14 Print Preview—Activity Usage Profile

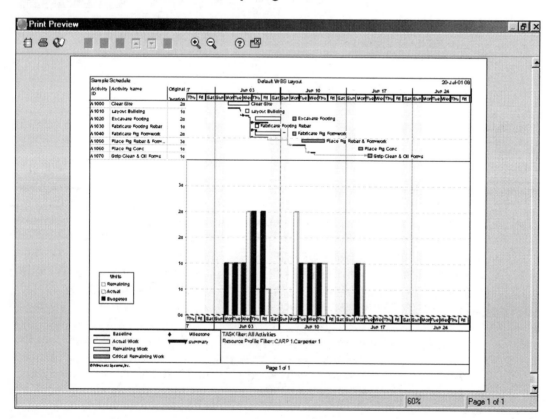

Figure 12-15 Print Preview—Resource Usage Profile

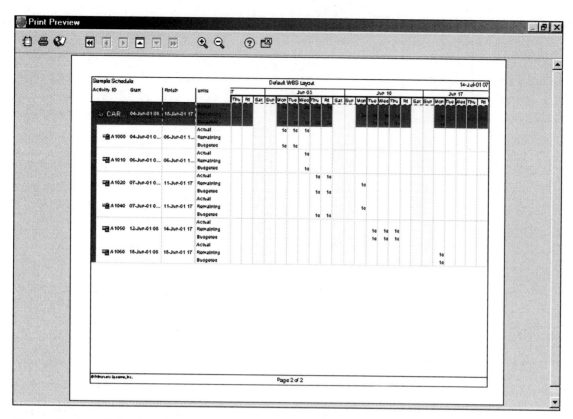

Figure 12–16   Print Preview—Resource Usage Spreadsheet

EXAMPLE PROBLEM

Figure 12–18 to Figure 12–21 show the update of resources of the schedule for the house put together as an example for student use (see the wood-framed house drawings in the Appendix). Figure 12–18 is the **Print Preview** of the **Resource Usage Spreadsheet** showing the updated resource information for the **CARPENTR** resource by activity. Figure 12–19 is the **Activity Usage Profile** hard-copy print showing the updated resource information for all resources. Figure 12–20 is the **Resource Usage Profile** hard-copy print showing the updated resource information for the **CARPENTR** resource for all activities. Figure 12–21 is the **Resource Control Report** hard-copy tabular print showing the updated resource information by resource for all activities.

Sample Schedule
Report Date 14-Jul-01 07

Start Date 04-Jun-01
Finish Date 20-Jun-01
Data Date 10-Jun-01

RC-01 Resource Control Report

User's Notes   **Student Constructors**

Resource

Activity ID	Cost Account	Budgeted Units	Units % Complete	Actual Unit	Remaining Unit	At Completio Unit
CARP 1	**Carpenter 1**					
A1000		2	100%	3	0	3
A1010		1	100%	1	0	1
A1020		2	50%	1	1	2
A1040		2	0%	0	1	1
A1050		3	0%	0	3	3
A1060		1	0%	0	1	1
Subtotal		**11d**	**45.45%**	**5d**	**6d**	**11d**
CARP 2	**Carpenter 2**					
A1010		1	100%	1	0	1
A1040		2	0%	0	1	1
Subtotal		**3d**	**50%**	**1d**	**1d**	**2d**
EQUIP OP	**Equipment Operator**					
A1020		2	50%	1	1	2
A1050		3	0%	0	3	3
A1060		1	0%	0	1	1
A1070		1	0%	0	1	1
Subtotal		**7d**	**15.38%**	**1d**	**6d**	**7d**
LAB 1	**Laborer 1**					
A1010		1	100%	2	0	2

Page 1 of 2

(c) Primavera Systems, Inc.

Figure 12–17   Resource Control Report

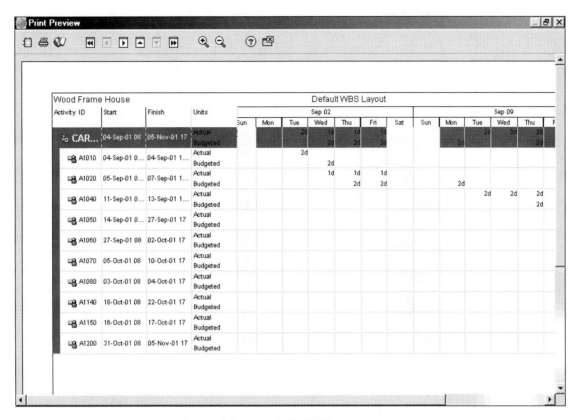

Figure 12–18   Resource Usage Spreadsheet—CARPENTR Resource

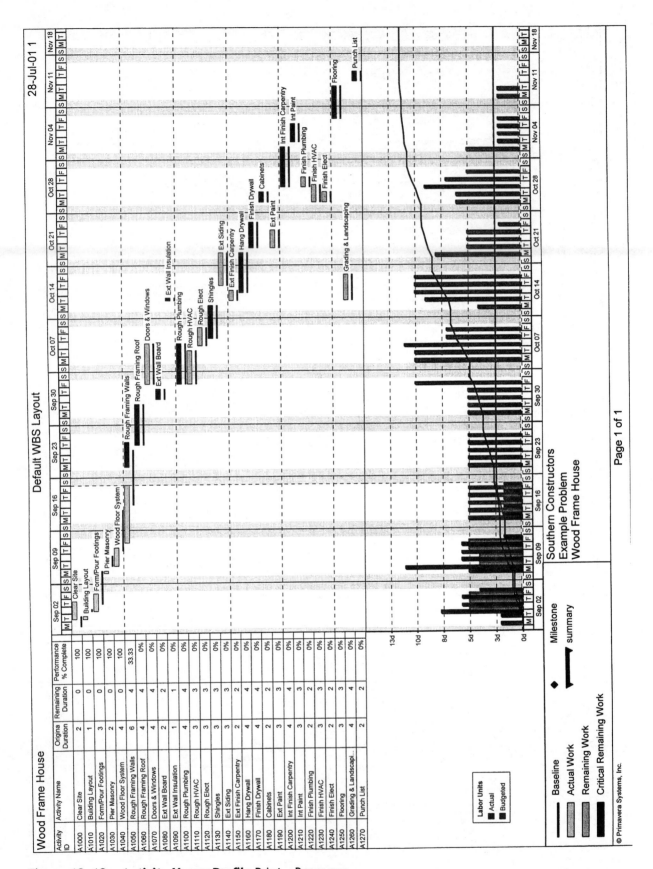

Figure 12–19 Activity Usage Profile Print—Resource

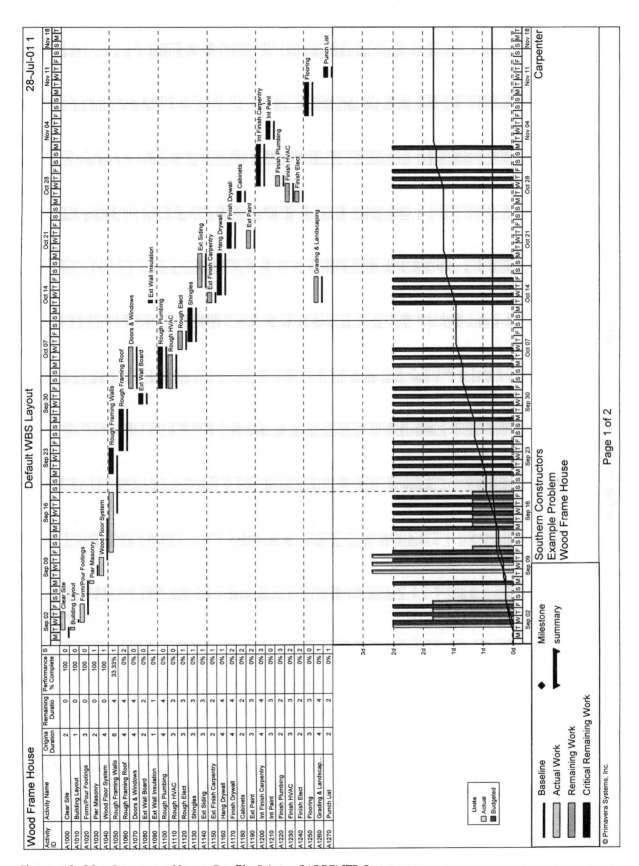

Figure 12–20   Resource Usage Profile Print—**CARPENTR** Resources

Wood Frame House
Report Date 28-Jul-01 17

RC-01 Resource Control Report

Start Date 04-Sep-01
Finish Date 15-Nov-01
Data Date 21-Sep-01

User's Notes   Student Constructors

Resource Activity ID	Cost Account	Budgeted Units	Units % Complete	Actual Unit	Remaining Unit	At Completio Unit
CARPENTR	Carpenter					
A1010		2	100%	2	0	2
A1020		6	100%	4	0	4
A1040		8	100%	7	0	7
A1050		12	33.33	4	8	12
A1060		8	0%	0	8	8
A1070		8	0%	0	8	8
A1080		4	0%	0	4	4
A1140		6	0%	0	6	6
A1150		4	0%	0	4	4
A1200		8	0%	0	8	8
Subtotal		66d	26.98%	17d	46d	63d
CARP FOR	Carpentry Forman					
A1010		1	100%	1	0	1
A1020		3	100%	3	0	3
A1040		4	100%	3	0	3
A1050		6	33.33	2	4	6
A1060		4	0%	0	4	4
A1070		4	0%	0	4	4
A1080		2	0%	0	2	2
A1140		3	0%	0	3	3
A1150		2	0%	0	2	2
A1200		4	0%	0	4	4

Page 1 of 5

(c) Primavera Systems, Inc.

Figure 12-21   Resource Control Tabular Report

EXERCISES

1. Produce a manual resource update (resource profile and table) for Figure 12–22.

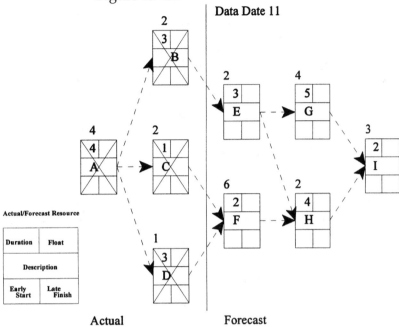

Figure 12–22   Exercise 1

2. Produce a manual resource update (resource profile and table) for Figure 12–23.

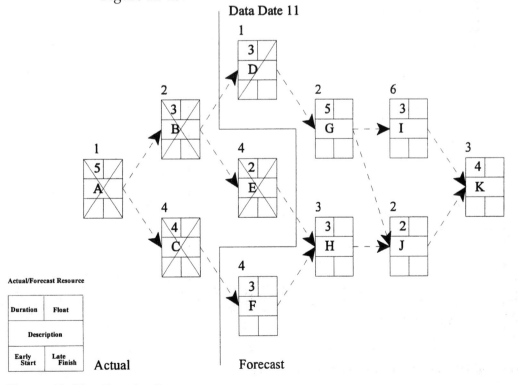

Figure 12–23   Exercise 2

3. Produce a manual resource update (resource profile and table) for Figure 12–24.

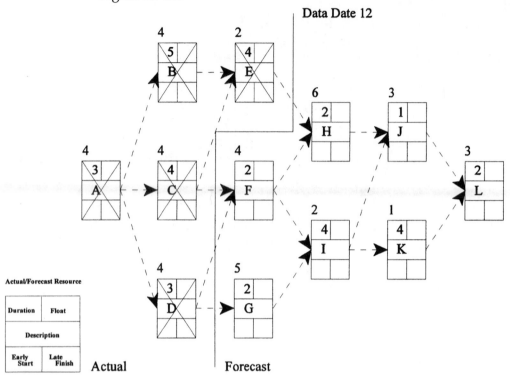

Figure 12–24 Exercise 3

4. Produce a manual resource update (resource profile and table) for Figure 12–25.

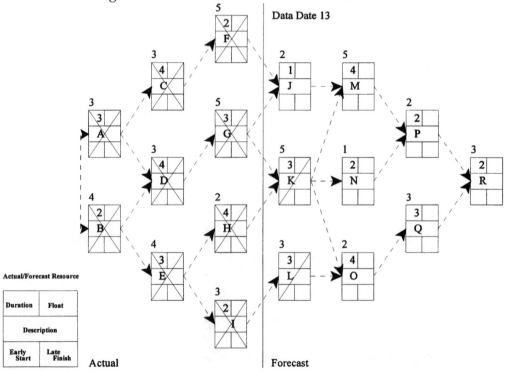

Figure 12–25 Exercise 4

5. Produce a manual resource update (resource profile and table) for Figure 12–26.

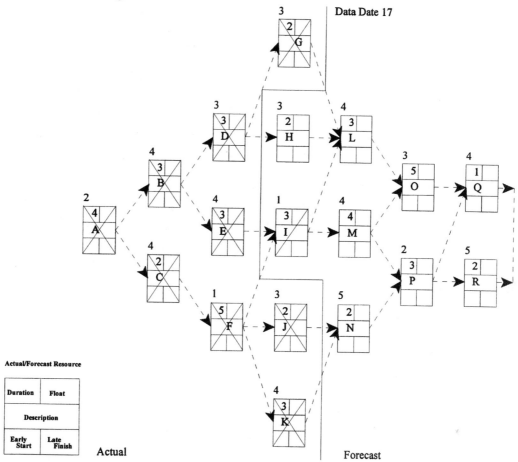

Figure 12–26   Exercise 5

6. Produce a manual resource update (resource profile and table) for Figure 12–27.

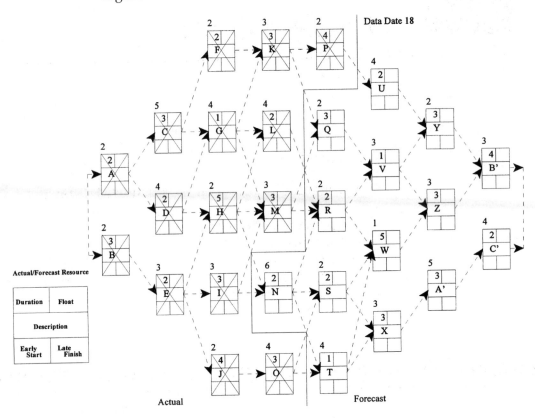

Figure 12–27 Exercise 6

7. Small Commercial Concrete Block Building—Resource Tracking
Prepare the following reports for the small commercial concrete block building located in the Appendix.
  a. Updated resource table.
  b. Updated resource profile.

8. Large Commercial Building—Resource Tracking
Prepare the following reports for the large commercial building located in the Appendix.
  a. Updated resource table.
  b. Updated resource profile.

13

Tracking Costs

Objectives

Upon completion of this chapter, you should be able to:

- Use updated cost tables
- Record expenditures
- Analyze current costs
- Use earned value costs
- Use updated cost profiles
- Print updated cost reports

TRACKING COSTS: COMPARING ACTUAL TO PLANNED EXPENDITURES

Once a project is under way, tracking costs and comparing actual to planned expenditures are tools in controlling costs. The costs can either be total costs or the accumulation of the resource costs of labor, equipment, materials, and/or other resources. These costs are usually assigned to the activity (allocated from the original estimate) when the baseline schedule is constructed. By tracking actual costs and comparing them to the baseline budget, you can determine cost progress and earned value.

Construction managers must determine whether the project is making or losing money. If there are problems, they determine where the problems lie; they then pass on the information to the right people in time to do something about it.

On a cost-plus or negotiated project, the owner, too, can use the cost-loaded and then cost-monitored schedule as valuable information in managing cash assets. This chapter shows how to use *P3e* to update costs to match the updated durations, logic, and activity changes made to the schedule in Chapter 11.

Cost Spreadsheet

The cost spreadsheet provides a convenient tool for the tracking and analysis of cost.

Figure 13–1 is a copy of the **Gantt Chart** view of the Sample Schedule showing the current and baseline schedules up top and the **Actual, Remaining,** and **Budgeted** expense costs in the **Activity Usage Spreadsheet** at the bottom of the screen. The current schedule was updated in Chapter 11, with progress, modified logic, and new activities input. The next step is to further update the schedule showing the actual expenditure of costs and to make forecasts.

The easiest way to evaluate costs on-screen is through the **Activity Usage Spreadsheet**. To obtain the **Activity Usage Spreadsheet**, click on the **Activity Usage Spreadsheet** button from the Activity Toolbar. The **Actual, Remaining,** and **Budgeted** expense cost rows show in the **Activity Usage Spreadsheet** in Figure 13–1. Note that, for the purposes of this chapter, the **Budgeted Expense Cost** is selected for comparison purposes rather than the **Baseline Expense Cost**. Either of these rows offers valuable cost information. To get the **Actual, Remaining,** and **Budgeted Expense Cost** rows to show, the **Activity Usage Spreadsheet** must be configured differently from the last time it was used in Chapter 12 to update resource information. To reconfigure the **Activity Usage Spreadsheet**, right click the mouse anywhere within the spreadsheet and select **Spreadsheet Fields** and the **Fields** dialog box will appear (Figure 13–2).

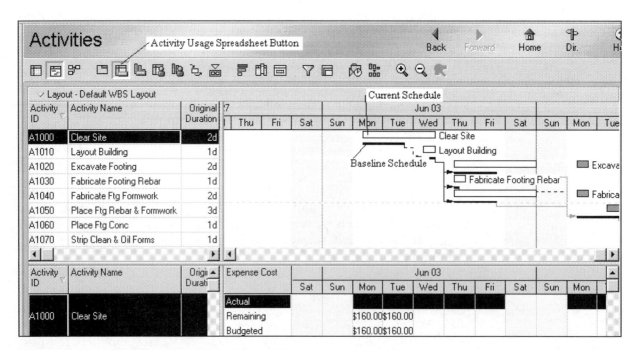

Figure 13–1   Activity Usage Spreadsheet Window

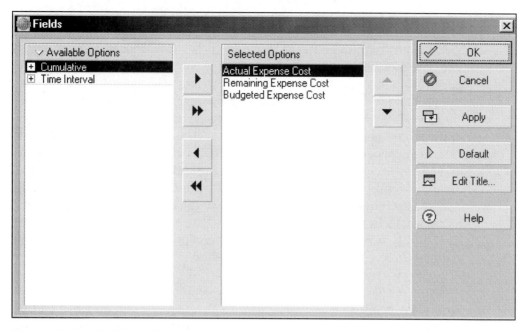

Figure 13–2   Fields Dialog Box

The **Fields** dialog box in Figure 13–2 is reconfigured to remove the resource information and include the **Actual, Remaining,** and **Budgeted Expense Cost** rows.

Budgets

Figure 13–1 is the current schedule, or the update of the baseline schedule as modified in Chapter 11. It contains the schedule and logic modifications made to update the schedule to the current data date. It also contains the actual input of resources made in Chapter 12. Now it is time to input the actual expense cost information to match the actual cost to date. The comparison of the budgeted costs to the actual costs of work performed makes cost forecasting possible. Cost forecasting is the projection of the future cost based on costs to date and management's knowledge of the project.

RECORDING EXPENDITURES

Costs must be gathered by activity. This is not the way most contractors' cost accounting systems work. The usual construction cost system gathers costs by cost account code. Labor time sheets, work measurement reports, purchase orders, and all other cost accounting documents are coded by cost account code. The estimate is organized in the same way. To be able to gather costs by activity means that another step must be added to the level of information gathered in the field. The documents must be coded with the cost accounting code and the activity ID for cost gathering purposes. Another element that makes the process of gathering costs by activity difficult is that indirect costs and profit are not typically placed in any specific activity but are prorated (or spread) over all activities. The same approach will be used in this chapter as was used in Chapter 7 where the baseline expense costs were inputted. Cost will be input only at the overall activity level and not broken down into labor, materials, equipment, overhead, and profit costs by activity. The ability to break down cost into detailed categories and cost codes is one of the real strengths of *P3e*. In the following section, we show how to record the actual cost expenditures for the Sample Schedule for activities where physical progress has been claimed during the updating of the baseline schedule.

Activity A1000

Select the **Expenses** tab from the **Activity Details** window (Figure 13–3). Click on the first activity to be updated with cost information, Activity

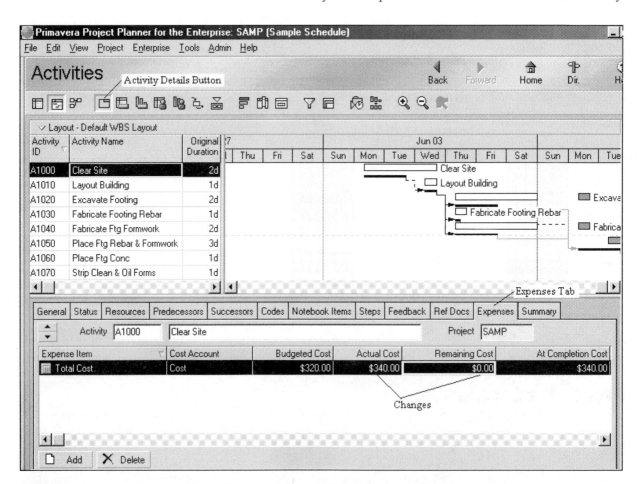

Figure 13–3　Activity Details Window—**Expenses** Tab—Activity A1000

A1000, Clear Site. If the columns in your **Expenses** tab of the **Activity Details** window do not appear as configured in Figure 13–3, you can reconfigure them. To reconfigure the columns of the **Expenses** tab, click the right mouse button while in the **Expenses** tab window and select **C̲ustomize Expenses Columns**. The **Expenses Columns** dialog box will appear for making column selections.

Budgeted Cost.   The $320.00 budget for Activity A1000, created when cost information for the baseline schedule was inputted in Chapter 7, is shown in Table 13–1. Note that the **Budgeted Expense Cost** row of the **Activity Usage Spreadsheet** (Figure 13–4) shows a budget of $160.00 to be expended on Activity A1000, for both Mon and Tue, for a total of $320.00.

As you can see in Table 13–1, the baseline expense cost by activity, the original duration for Activity A1000 was 2 days, or $320/2 days or $160 per day. When progress is updated to create the current schedule, the actual duration for Activity A1000 was 3-days. From Table 13–2, the cost update to current schedule, the $320 budget is now an actual coat of $340 or $113.33/day.

Actual Cost.   As can be seen from Figure 13–3, the **Actual Cost** column of the **Expenses** tab for Activity A1000 has been changed to reflect the total actual cost of $340.00. The $340.00 actual total cost for Activity A1000 was determined when actual cost information for the activity was gathered (Table 13–2). Note in Figure 13–4, the **Activity**

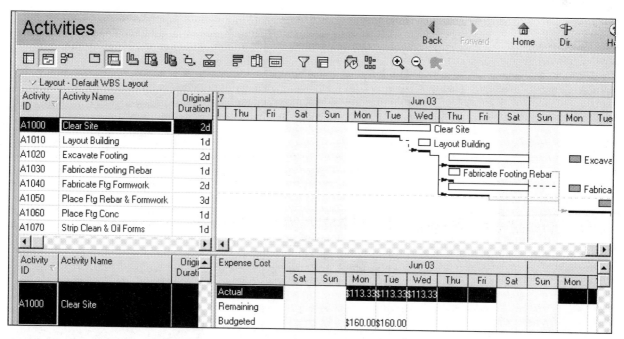

Figure 13–4   Activity Usage Spreadsheet Window—Activity A1000

Activity		Duration	Day 1 Mon	Day 2 Tue	Day 3 Wed	Day 4 Thu	Day 5 Fri	Day 6 Mon	Day 7 Tue	Day 8 Wed
A1000	Clear Site	2	$160	$160						
A1010	Building Layout	1			$346					
A1020	Excavate Footings	2				$370	$370			
A1030	Fabricate Ftg Rebar	1				$198				
A1040	Fabricate Ftg Formwork	2				$473	$473			
A1050	Place Ftg Rebar & Formwork	2						$392	$392	
A1060	Place Ftg Conc	1								$892
	Cost/Day		$160	$160	$346	$1,041	$843	$392	$392	$892
	Cumulative Cost/Day		$160	$320	$666	$1,707	$2,550	$2,942	$3,334	$4,226

Table 13–1 Baseline Expense Cost

Data Date

Activity		Duration	Day 1 Mon	Day 2 Tue	Day 3 Wed	Day 4 Thu	Day 5 Fri	Day 6 Mon	Day 7 Tue	Day 8 Wed	Day 9 Thu	Day 10 Fri	Day 11 Mon	Day 12 Tue
A1000	Clear Site	3	$113	$113	$113									
A1010	Building Layout	1			$372									
A1020	Excavate Footings	3				$185	$185	$370						
A1030	Fabricate Ftg Rebar	1				$198								
A1040	Fabricate Ftg Formwork	3				$237	$237	$598						
A1050	Place Ftg Rebar & Formwork	3							$359	$359	$359			
A1060	Place Ftg Conc	1											$892	
A1070	Strip Clean & Oil Forms	1												$200
	Cost/Day		$113	$113	$485	$620	$422	$968	$359	$359	$359	$0	$892	$200
	Cumulative Cost/Day		$113	$226	$711	$1,331	$1,753	$2,721	$3,080	$3,439	$3,798	$3,798	$4,690	$4,890

Table 13–2 Actual/Forecast Cost Input

Usage Spreadsheet, that the **Actual Expense Cost** row for Activity A1000 now shows actual costs of $113.33 expended on Mon, Tue, and Wed, for a total activity cost of $340.00.

Remaining Cost.   Even though the activity **Actual Cost** has been input, and the activity is updated at 100% complete, you must still change the **Remaining Cost** field (Figure 13–3) to reflect $0.00 left to be spent on this activity. Note in Figure 13–4 that the **Remaining Expense Cost** row of the **Activity Usage Spreadsheet** for Activity A1000 now shows no remaining cost to be expended.

At Completion Cost.   The **At Completion Cost** is the sum of the **Actual Cost** field plus the **Remaining Cost** fields. The **At Completion Cost** field for Activity A1000 (Figure 13–3) shows $340.00. The $340.00 in the **At Completion Cost** field reflects the $340.00 **Actual Cost** field plus the $0.00 **Remaining Cost** field.

Activity A1010

Click on the next activity to be updated. Activity A1010, Layout Building, is selected (Figure 13–5) while still in the **Expenses** tab of the **Activity Details** window.

Budgeted Cost.   The **Budgeted Cost** column of the **Expenses** tab for Activity A1010 shows $346.00 (Figure 13–5). The $346.00 budget for Activity A1010 is shown in Table 13–1. Note that the **Budgeted Expense Cost** row of the **Activity Usage Spreadsheet** (Figure 13–6) shows a budget of $346.00, for Activity A1010, to be expended on Wed.

Actual Cost.   As can be seen from Figure 13–5, the **Actual Cost** column of the **Expenses** tab for Activity A1010 has been changed to reflect the total actual cost of $372.00. The $372.00 actual total cost for Activity A1010 was determined when actual cost information for the activity was gathered (Table 13–2). Note that in, the **Activity Usage Spreadsheet** of Figure 13–6, the **Actual Expense Cost** row for Activity A1010 now shows an actual cost of $372.00 expended on Wed.

Remaining Cost.   Even though the activity **Actual Cost** has been input, and the activity is updated at 100% complete, you must still change the **Remaining Cost** field (Figure 13–5) to reflect $0.00 left to be spent on this activity. Note that in Figure 13–6 the **Remaining Expense Cost** row of the **Activity Usage Spreadsheet** for Activity A1010 now shows no remaining cost to be expended.

At Completion Cost.   The **At Completion Cost** is the sum of the **Actual Cost** field plus the **Remaining Cost** field. The **At Completion**

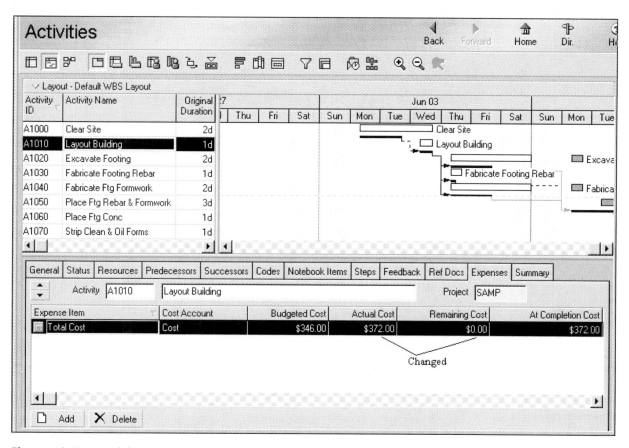

Figure 13–5   Activity Details Window—**Expenses** Tab—Activity A1010

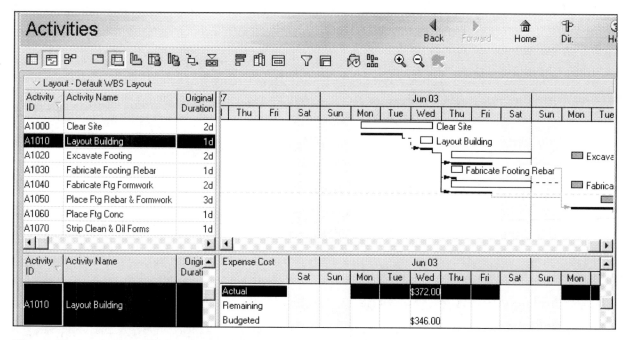

Figure 13–6   Activity Usage Spreadsheet Window—Activity A1010

Cost field for Activity A1010, Figure 13–5, shows $372.00. The **At Completion Cost** of $372.00 reflects the $372.00 **Actual Cost** plus the $0.00 **Remaining Cost**.

Activity A1020

Click on the next activity to be updated. Activity A1020, Excavate Footings, is selected (Figure 13–7) while still in the **Expenses** tab of the **Activity Details** window.

Budgeted Cost.   The **Budgeted Cost** column of the **Expenses** tab for Activity A1020 shows $739.00 (Figure 13–7). The $739.00 budget for Activity A1020 was created when cost information for the baseline schedule was input (shown in Table 13–1). Note that the **Budgeted Expense Cost** row of the **Activity Usage Spreadsheet** (Figure 13–8) shows a budget of $739.00 for Activity A1020. The $739 budget was to be expended at $369.50 for both Thu and Fri.

Actual Cost.   As can be seen from Figure 13–7, the **Actual Cost** column for Activity A1020 has been changed to reflect an actual cost expended of $370.00. The $370.00 actual total cost for Activity A1020 is shown in Table 13–2. Note in Figure 13–8 that the **Actual Expense Cost** row for Activity A1020 now shows actual costs of $185.00 expended on Thu and Fri for a total of $370.00.

Remaining Cost.   The **Remaining Cost** field (Figure 13–7) was changed to reflect $370.00 left to be spent on this activity (Table 13–2). Note in Figure 13–8 that the **Remaining Expense Cost** row for Activity A1020 now shows $370.00 remaining to be spent.

At Completion Cost.   The **At Completion Cost** is the sum of the **Actual Cost** field plus the **Remaining Cost** field. The **At Completion Cost** field for Activity A1020, Figure 13–7, shows $740.00. The $740.00 **At Completion Cost** reflects the $370.00 **Actual Cost** plus the $370.00 **Remaining Cost**.

Activities A1030 and A1040

Activity A1030, Fabricate Footing Rebar, and Activity A1040, Fabricate Footing Formwork, of the Sample Schedule were cost updated using the same procedures used for Activities A1000, A1010, and A1020.

Activity A1050

Click on the next activity to be updated. Activity A1050, Place Ftg Rebar & Formwork, is selected (Figure 13–9) while in the **Expenses** tab of the

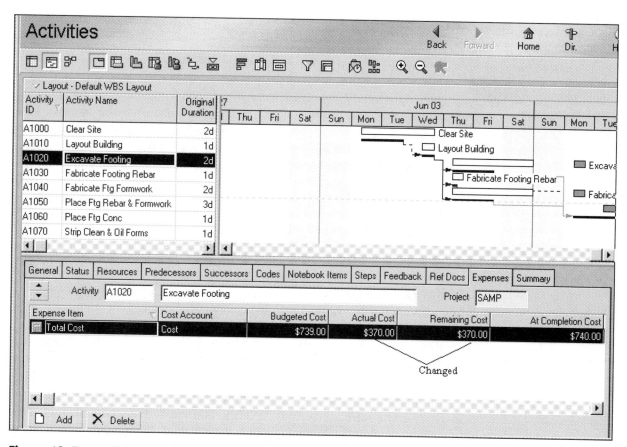

Figure 13–7   Activity Details Window—**Expenses** Tab—Activity A1020

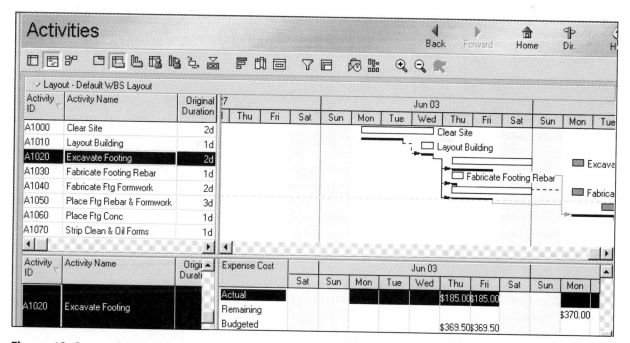

Figure 13–8   Activity Usage Spreadsheet Window—Activity A1020

Activity Details window.

> **Budgeted Cost.** A $784.00 budget for Activity A1050 was created when cost information for the baseline schedule was inputted as is shown in Table 13–1. The $1,077.00 revised budgeted cost for Activity A1050 was determined when cost information for the activity was gathered (Table 13–2). The **Budgeted Cost** column of the **Expenses** tab for Activity A1050 had been changed to reflect $1077.00 (Figure 13–9). Note that the **Budgeted Expense Cost** row of the **Activity Usage Spreadsheet** (Figure 13–10) shows a budget of $1077.00 for Activity A1050 to be expended at $359.00 each day on Tue, Wed, and Thu.
>
> **Actual Cost.** As can be seen from Figure 13–9, the **Actual Cost** column of the **Expenses** tab for Activity A1050 has been left at $0.00 expended since there has been no progress. Note in Figure 13–10 that the **Actual Expense Cost** row of the **Activity Usage Spreadsheet** for Activity A1050 shows that no actual costs have been expended.
>
> **Remaining Cost.** With the **Budgeted Cost** set at $1077.00 and the **Actual Cost** set at $0.00, *P3e* automatically puts the **Remaining Cost** field at $1077.00 to be spent on this activity (Table 13–2). Note in Figure 13–9 that the **Remaining Expense Cost** row of the **Activity Usage Spreadsheet** for Activity A1050 now shows $1077.00 remaining to be spent. The $1077.00 is to be split, with $359.00 being spent each day on Tue, Wed, and Thu.
>
> **At Completion Cost.** The **At Completion Cost** (Figure 13–9) is the same as the **Budgeted Cost** since no progress has been achieved on this activity. The **At Completion Cost** field (Figure 13–9) for Activity A1050 shows $1,077.00. The $1,077.00 **At Completion Cost** reflects the $1,077.00 **Budgeted Cost**.

Activity A1060

Activity A1060, Place Ftg Conc, has no progress or changes to its budget. Therefore no changes to the cost information for Activity A1060 are necessary.

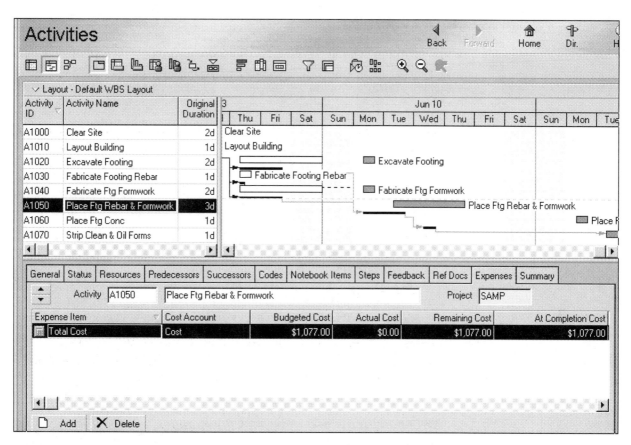

Figure 13–9   Activity Details Window—**Expenses** Tab—Activity A1050

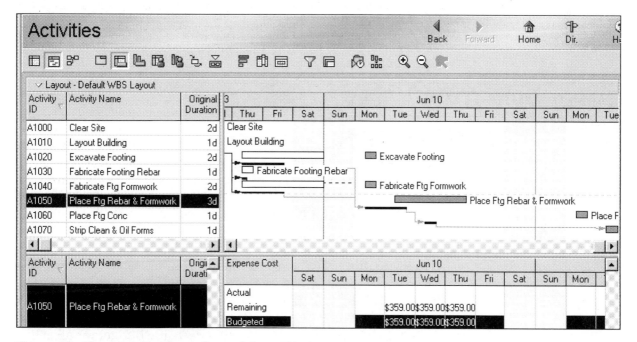

Figure 13–10   Activity Usage Spreadsheet Window—Activity A1050

Activity A1070

Click on the next activity to be updated. Activity A1070, Strip Clean & Oil Forms, is selected (Figure 13–11) while still in the **Expenses** tab of the **Activity Details** window.

Budgeted Cost.   A budget for Activity A1070 was not created when cost information for the baseline schedule was input in Chapter 7 (Table 13–1). Activity A1070 was added when the Sample Schedule was updated in Chapter 11, but the budget was never changed to reflect the added cost of this activity. Click on the **Add** button of the **Expenses** tab (Figure 13–11), to add cost information. In Figure 13–12, the **Budgeted Cost** field for Activity A1070 has been changed to create a $200.00 budget (Table 13–2).

Actual Cost.   As can be seen from Figure 13–12, the **Actual Cost** column of the **Expenses** tab for Activity A1070 has been left at $0.00 expended since there has been no progress achieved on this activity.

Remaining Cost.   With the **Budgeted Cost** set at $200.00 (Figure 13–12) and the **Actual Cost** set at $0.00, *P3e* automatically puts the **Remaining Cost** field at $200.00 to be spent on this activity.

At Completion Cost.   The **At Completion Cost** is the same as the **Budgeted Cost** since no progress has been achieved on this activity. The **At Completion Cost** field (Figure 13–12) for Activity A1070 shows $200.00. The $200.00 **At Completion Cost** reflects the $200.00 **Budgeted Cost**.

EARNED VALUE—TABULAR REPORTS

P3e provides some very beneficial tabular reports for cost analysis and forecasting. One example is the **AC-01 Activity Earned Value** report that can be accessed from the **Reports** selection from the **Tools** main pull-down menu. The columns provided in the **AC-01 Activity Earned Value** report for cost analysis are: **Activity Status, BCWS, BCWP, ACW, BAC, ETC, EA,** and **VA.**

- **Activity Status.**   The **Activity Status** column lets you know at a glance whether the activity is Completed, In Progress, or Not started.
- **BCWS.**   The **BCWS** (Budgeted Cost of Work Scheduled) lets you know the baseline budget for the work performed to date. In Figure 13–13, the **BCWS** value of $2,496.00 reflects the budget for all activities that have been started.

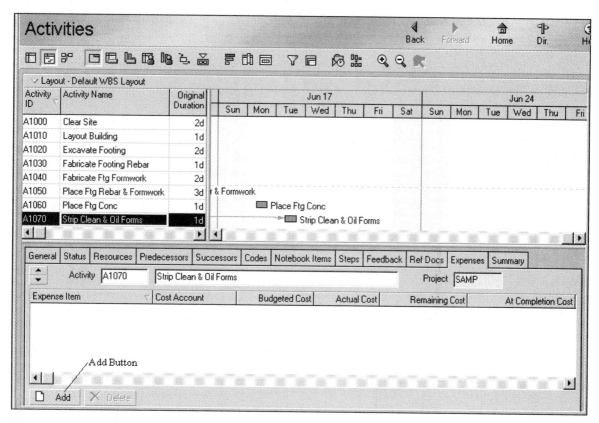

Figure 13–11   Activity Details Window—**Expenses** Tab—Activity A1070

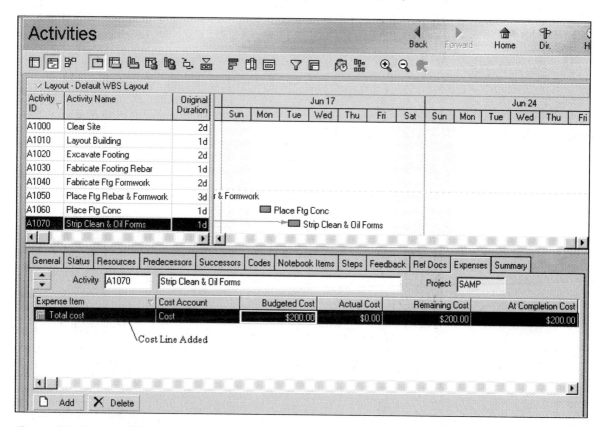

Figure 13–12   Activity Details Window—**Expenses** Tab—Activity A1070 with Cost Added

Sample Schedule
Report Date 23-Jul-01 08

Project Start 04-Jun-01
Project Finish 19-Jun-01
Data Date 10-Jun-0

AC-01 Activity Earned Value

WBS

Activity ID	Activity Name	Activity Status	BCWS	BCWP	ACW	BAC	ETC	EA	VA
SAMP	**Sample Schedule**								
A1000	Clear Site	Completed	$320.00	$320.00	$340.00	$320.00	$0.0	$340.00	($20.00)
A1010	Layout Building	Completed	$346.00	$346.00	$372.00	$346.00	$0.0	$372.00	($26.00)
A1020	Excavate Footing	In Progress	$739.00	$369.50	$370.00	$739.00	$370.00	$740.00	($1.00)
A1030	Fabricate Footing Rebar	Completed	$144.00	$144.00	$198.00	$144.00	$0.0	$198.00	($54.00)
A1040	Fabricate Ftg Formwork	In Progress	$947.00	$473.50	$474.00	$947.00	$598.00	$1,072.00	($125.00)
A1050	Place Ftg Rebar & Formwork	Not Started	$0.0	$0.0	$0.0	$784.00	$1,077.00	$1,077.00	($293.00)
A1060	Place Ftg Conc	Not Started	$0.0	$0.0	$0.0	$892.00	$892.00	$892.00	$0.0
A1070	Strip Clean & Oil Forms	Not Started	$0.0	$0.0	$0.0	$0.0	$200.00	$200.00	($200.00)
Subtotal			$2,496.00	$1,653.00	$1,754.00	$4,172.00	$3,137.00	$4,891.00	($719.00)
Total			$2,496.00	$1,653.00	$1,754.00	$4,172.00	$3,137.00	$4,891.00	($719.00)

Page 1 of 1

© Primavera Systems, Inc.

Figure 13–13 Activity Earned Value Report

- **BCWP.** The **BCWP** (Budgeted Cost of Work Performed or Earned Value) lets you know how much of the baseline budget has been earned by actual progress. In Figure 13–13, the **BCWP** value of $1,653.00 reflects that portion of the baseline budget that has been earned by work put in place.

- **ACWP.** The **ACWP** (Actual Cost of Work Performed) lets you know the actual cost of work put in place. In Figure 13–13, the **ACWP** value of $1,754.00 reflects the cost of work put in place. The Actual Cost of Work Performed (**ACWP**) value of $1,754.00 minus the Budgeted Cost of Work Performed or Earned Value (**BCWP**) of $1,653.00 results in a loss of $101.00 for the work performed on activities to date.

- **BAC.** The **BAC** (Budgeted at Completion) lets you know the baseline budget for all work. In Figure 13–13, the **BAC** value of $4,172.00 reflects the original baseline budget for all activities.

- **ETC.** The **ETC** (Estimate to Completion) lets you know the forecast changes to the baseline budget for all work still to be completed. In Figure 13–13, the **ETC** value of $3,137.00 reflects the baseline budget for all work still to be completed. Note the changes in Activities A1050 and A1070. The budgets of these two activities were changed from their baseline value when the Sample Schedule was updated with the current cost forecast.

- **EAC.** The **EAC** (Estimate at Completion) lets you know the forecast of final project cost at completion. In Figure 13–13, the **EAC** value of $4,891.00 reflects the forecast of final project cost at completion. Note that the **EAC** (Estimate at Completion) value of $4,891.00 minus the **BAC** (Budgeted at Completion) value of $4,172.00 means there is a $719.00 difference between the baseline budget and the estimate to complete. The project is possibly over budget by this amount. Note the major changes that occur in Activities A1050 and A1070 where the budgets were changed from the baseline when the Sample Schedule was updated with the current cost forecast.

- **VAC.** The **VAC** (Variance at Completion) lets you know the difference between the **EAC** (Estimate at Completion) and the **BAC** (Budgeted at Completion). In Figure 13–13, the **VAC** value of $719.00 reflects the difference between the baseline budget and the estimate to complete. This value reflects the possible loss on the project. Hopefully this amount can be recovered through a possible change order or contract modification to reflect the additional work.

It is important to highlight the fact that only **BCWP** (Budgeted Cost of Work Performed) and **ACWP** (Actual Cost of Work Performed) should be compared for cost purposes, and only **BCWP** (Budgeted Cost of Work Performed) and **BCWS** (Budgeted Cost of Work Scheduled) should be compared for schedule purposes. Any comparison made between **BCWS** and **ACWP** will be very misleading for technical reasons.

COST PROFILE

So far in this chapter, cost information has been presented in an **Activity Usage Spreadsheet** window (Figure 13–10) in the on-screen representation. Sometimes a cost profile or graphical representation is more helpful. Click on the **Activity Usage Profile** button of the Activity Toolbar and the **Activity Usage Profile** will appear at the bottom of the screen (Figure 13–14). Since the last time we used this profile was in Chapter 12, updating resources, it must be reconfigured for cost. While in the **Activity Usage Profile** window, click the right mouse button and select **Activity Usage Profile Options**; the **Activity Usage Profile Options** dialog box will appear (Figure 13–15). This dialog box is used to reconfigure the **Activity Usage Profile** in Figure 13–14 to show the **Cost**. The costs displayed are filtered to show only **Expenses**. The checkboxes for the **Budgeted, Actual**, and **Remaining** costs are selected for display. The checkboxes to show the **Cumulative** cost curves for the **Budgeted, Actual,** and **Remaining** costs are also selected for display. The **Activity Usage Profile** in Figure 13–14 is a result of the **Activity Usage Profile Options** dialog box options selected in Figure 13–15.

Figure 13–16 is the **Print Preview** of the Sample Schedule configured to show the **Activity Usage Profile**.

EXAMPLE PROBLEM

Figure 13–17 and Figure 13–18 show the cost update for the house put together as an example for student use (see the wood-framed house drawings in the Appendix). Figure 13–17 is a hard-copy print of the updated *P3e* Gantt chart with the **Activity Usage Profile** shown. Figure 13–18 is a hard-copy print of the updated *P3e* **Activity Earned Value Report**.

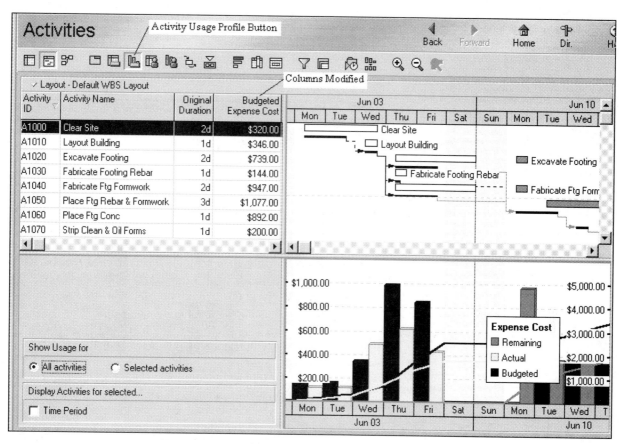

Figure 13–14   Activity Usage Profile Window Showing Cost

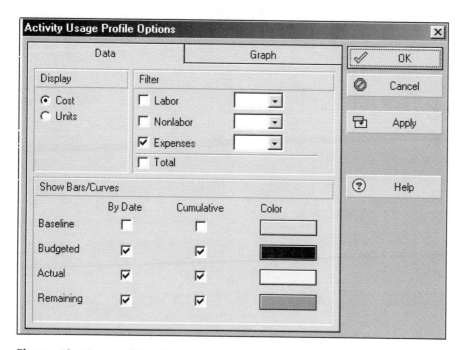

Figure 13–15   Activity Usage Profile Options Dialog Box

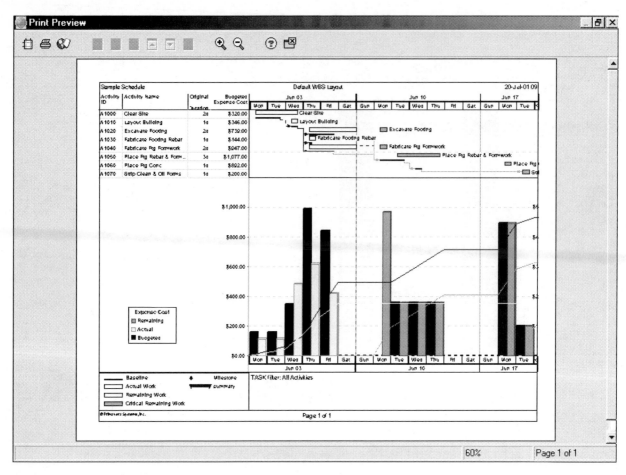

Figure 13–16   Print Preview of **Activity Usage Profile** Showing Cost

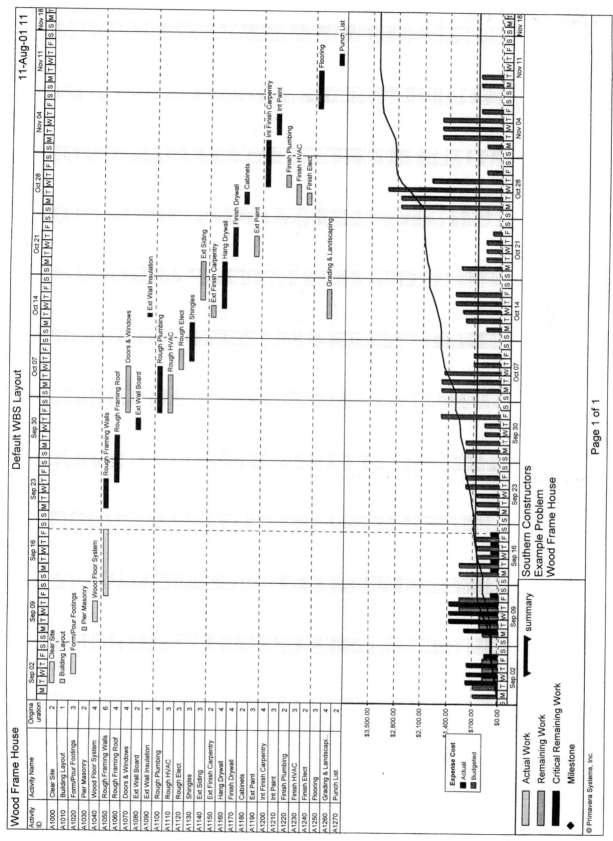

Figure 13–17 Activity Usage Profile Print—Costs

Wood Frame House
Report Date 29-Jul-01 06

Project Start 04-Sep-01
Project Finish 15-Nov-01
Data Date 21-Sep-01

AC-01 Activity Earned Value

WBS

Example Wood Frame House

Activity ID	Activity Name	Activity Status	BCWS	BCWP	ACWP	BAC	ETC	EAC	VAC
A1000	Clear Site	Completed	$1,280.00	$1,280.00	$1,440.00	$1,280.00	$0.00	$1,440.00	($160.00)
A1010	Building Layout	Completed	$386.00	$386.00	$370.00	$386.00	$0.00	$370.00	$16.00
A1020	Form/Pour Footings	Completed	$1,174.00	$1,174.00	$1,050.00	$1,174.00	$0.00	$1,050.00	$124.00
A1030	Pier Masonry	Completed	$967.00	$967.00	$910.00	$967.00	$0.00	$910.00	$57.00
A1040	Wood Floor System	Completed	$4,181.00	$4,181.00	$3,950.00	$4,181.00	$0.00	$3,950.00	$231.00
A1050	Rough Framing Walls	In Progress	$1,661.50	$1,107.67	$1,100.00	$3,323.00	$2,223.00	$3,323.00	$0.00
A1060	Rough Framing Roof	Not Started	$0.00	$0.00	$0.00	$3,468.00	$3,468.00	$3,468.00	$0.00
A1070	Doors & Windows	Not Started	$0.00	$0.00	$0.00	$3,995.00	$3,995.00	$3,995.00	$0.00
A1080	Ext Wall Board	Not Started	$0.00	$0.00	$0.00	$736.00	$736.00	$736.00	$0.00
A1090	Ext Wall Insulation	Not Started	$0.00	$0.00	$0.00	$385.00	$385.00	$385.00	$0.00
A1100	Rough Plumbing	Not Started	$0.00	$0.00	$0.00	$750.00	$750.00	$750.00	$0.00
A1110	Rough HVAC	Not Started	$0.00	$0.00	$0.00	$1,168.00	$1,168.00	$1,168.00	$0.00
A1120	Rough Elect	Not Started	$0.00	$0.00	$0.00	$940.00	$940.00	$940.00	$0.00
A1130	Shingles	Not Started	$0.00	$0.00	$0.00	$1,091.00	$1,091.00	$1,091.00	$0.00
A1140	Ext Siding	Not Started	$0.00	$0.00	$0.00	$1,710.00	$1,710.00	$1,710.00	$0.00
A1150	Ext Finish Carpentry	Not Started	$0.00	$0.00	$0.00	$736.00	$736.00	$736.00	$0.00
A1160	Hang Drywall	Not Started	$0.00	$0.00	$0.00	$1,844.00	$1,844.00	$1,844.00	$0.00
A1170	Finish Drywall	Not Started	$0.00	$0.00	$0.00	$790.00	$790.00	$790.00	$0.00
A1180	Cabinets	Not Started	$0.00	$0.00	$0.00	$1,618.00	$1,618.00	$1,618.00	$0.00
A1190	Ext Paint	Not Started	$0.00	$0.00	$0.00	$525.00	$525.00	$525.00	$0.00
A1200	Int Finish Carpentry	Not Started	$0.00	$0.00	$0.00	$1,472.00	$1,472.00	$1,472.00	$0.00
A1210	Int Paint	Not Started	$0.00	$0.00	$0.00	$4,725.00	$4,725.00	$4,725.00	$0.00
A1220	Finish Plumbing	Not Started	$0.00	$0.00	$0.00	$3,000.00	$3,000.00	$3,000.00	$0.00

Page 1 of 2

Figure 13–18a   Activity Earned Value Report

Wood Frame House
Report Date 29-Jul-01 06

Project Start 04-Sep-01
Project Finish 15-Nov-01
Data Date 21-Sep-01

AC-01 Activity Earned Value

WBS

Activity ID	Activity Name	Activity Status	BCWS	BCWP	ACWP	BAC	ETC	EAC	VAC
A1230	Finish HVAC	Not Started	$0.00	$0.00	$0.00	$3,506.00	$3,506.00	$3,506.00	$0.00
A1240	Finish Elect	Not Started	$0.00	$0.00	$0.00	$1,410.00	$1,410.00	$1,410.00	$0.00
A1250	Flooring	Not Started	$0.00	$0.00	$0.00	$1,583.00	$1,583.00	$1,583.00	$0.00
A1260	Grading & Landscaping	Not Started	$0.00	$0.00	$0.00	$600.00	$600.00	$600.00	$0.00
A1270	Punch List	Not Started	$0.00	$0.00	$0.00	$0.00	$0.00	$0.00	$0.00
Subtotal			$9,649.50	$9,095.67	$8,820.00	$47,363.00	$38,275.00	$47,095.00	$268.00
Total			$9,649.50	$9,095.67	$8,820.00	$47,363.00	$38,275.00	$47,095.00	$268.00

Page 2 of 2

© Primavera Systems, Inc.

Figure 13–18b   Activity Earned Value Report

EXERCISES

1. Produce a manual resource update (resource profile and table) for Figure 13–19.

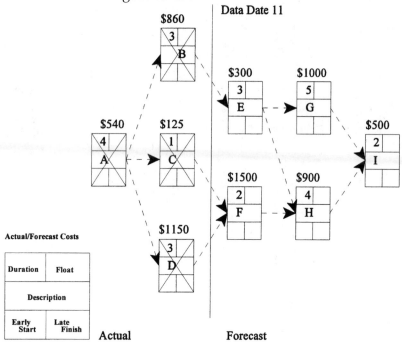

Figure 13–19   Exercise 1

2. Produce a manual resource update (resource profile and table) for Figure 13–20.

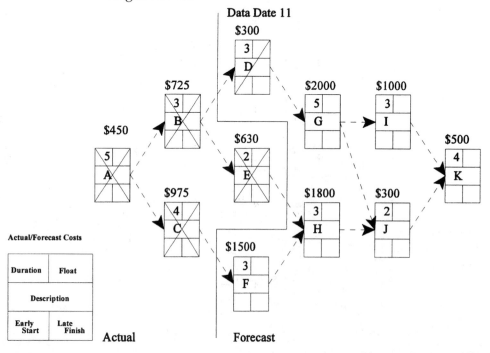

Figure 13–20   Exercise 2

3. Produce a manual resource update (resource profile and table) for Figure 13–21.

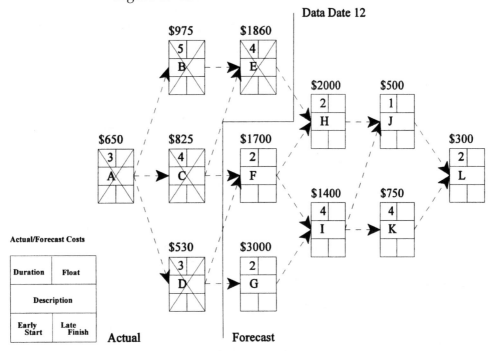

Figure 13–21   Exercise 3

4. Produce a manual resource update (resource profile and table) for Figure 13–22.

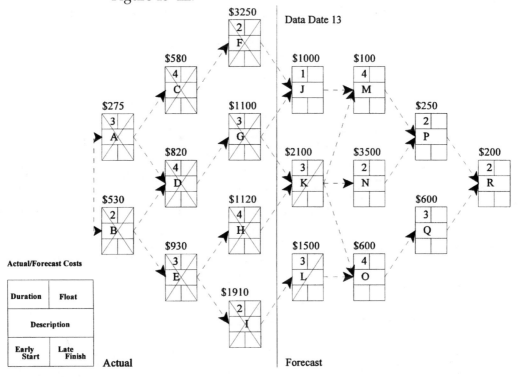

Figure 13–22   Exercise 4

5. Produce a manual resource update (resource profile and table) for Figure 13–23.

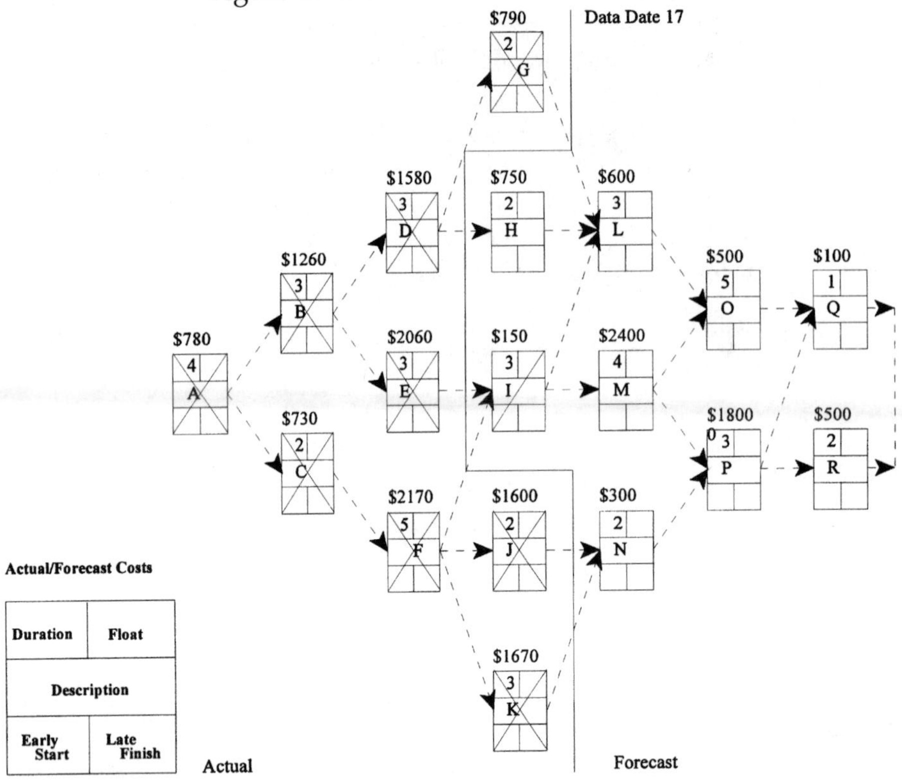

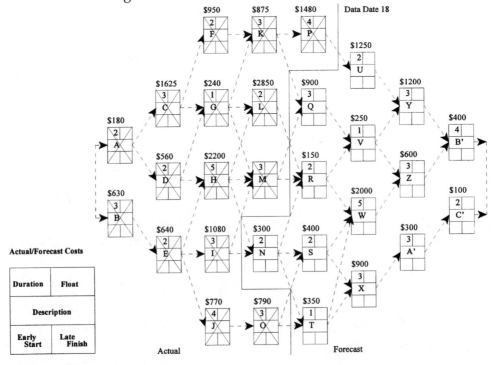

Figure 13–23   Exercise 5

6. Produce a manual resource update (resource profile and table) for Figure 13–24.

Figure 13–24   Exercise 6

7. **Small Commercial Concrete Block Building—Resource Tracking**
   Prepare the following reports for the small commercial concrete block building located in the Appendix.
   A. Updated earned value report.
   B. Updated cost profile.

8. **Large Commercial Building—Resource Tracking**
   Prepare the following reports for the large commercial building located in the Appendix.
   A. Updated earned value report.
   B. Updated cost profile.

Appendix

Drawings for Example Problems and Exercises

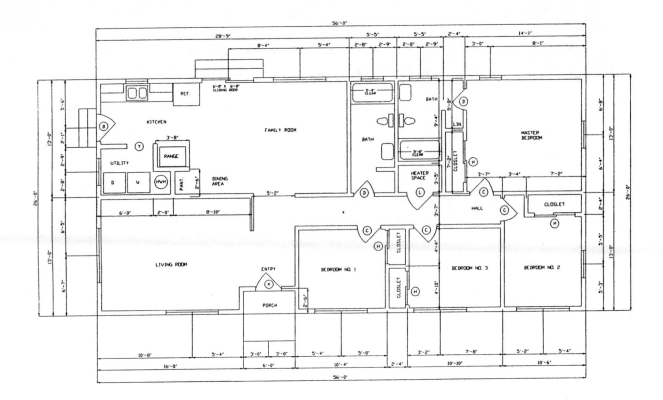

FLOOR PLAN
SCALE : 1/4" ==== 1'-0"

COMPOSITION SHINGLES

VINYL SIDING

ELEVATION
SCALE : 1/4" ==== 1'-0"

Figure A–1 Wood-Framed House—Floor Plan

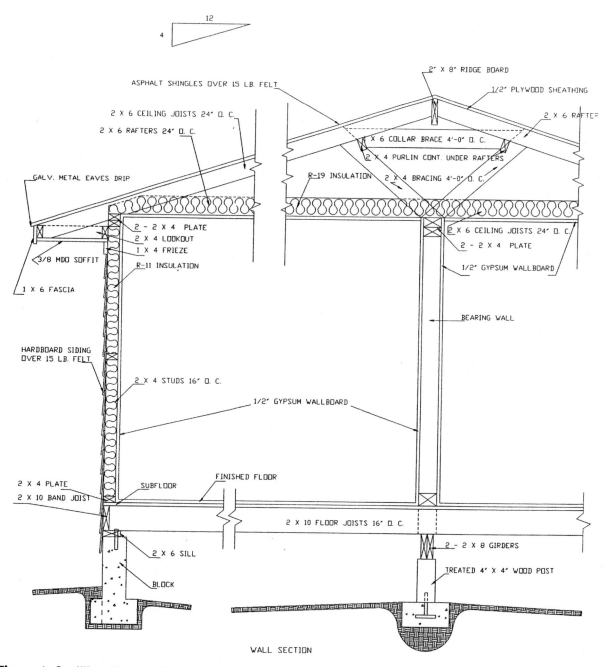

Figure A–2 Wood-Framed House—Section

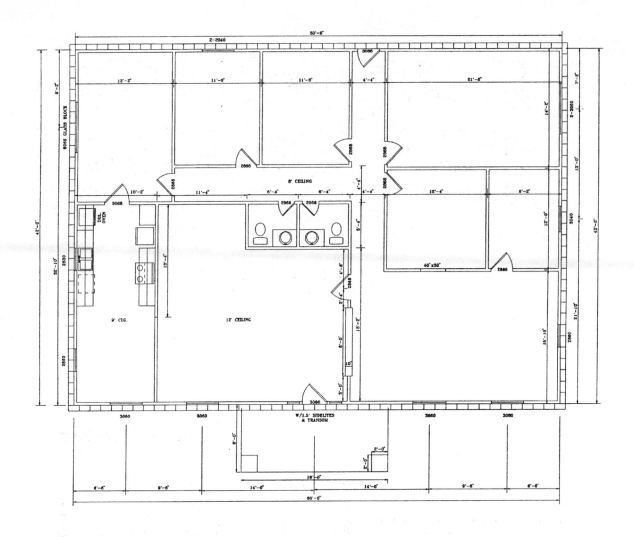

FLOOR PLAN
SCALE : 1/4" ==== 1'-0"

Figure A–3 Small Commercial Concrete Block Building—Floor Plan

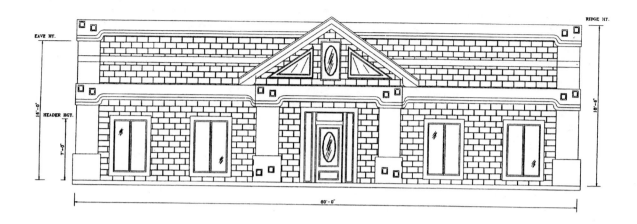

FRONT ELEVATION

SCALE : 1/4" ==== 1'-0"

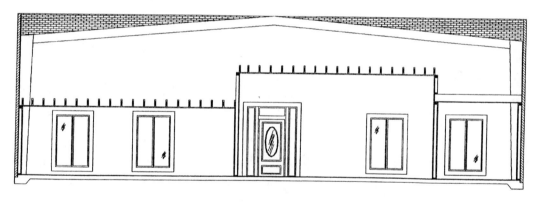

BUILDING SECTION

SCALE : 1/4" ==== 1'-0"

Figure A–4   Small Commercial Concrete Block Building—Section

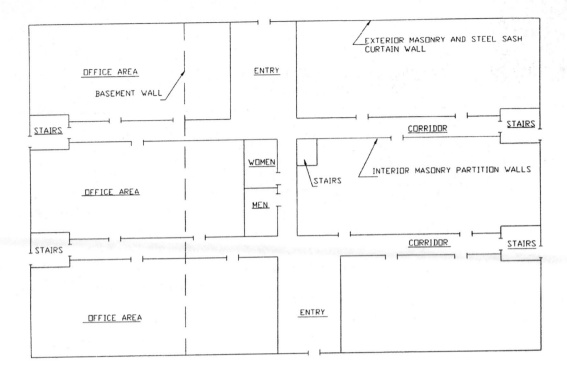

FLOOR PLAN

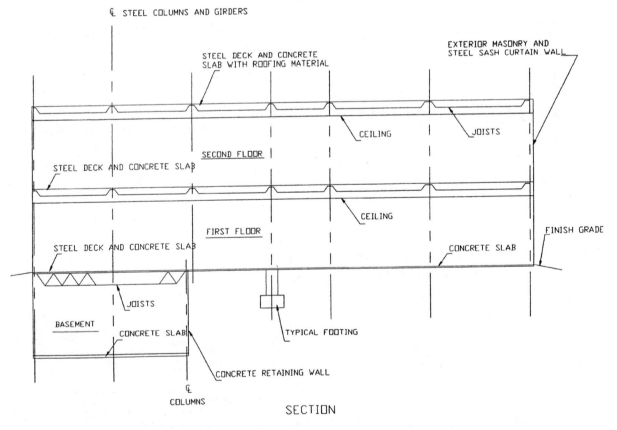

SECTION

Figure A-5 Large Commercial Building—Floor Plan

Index